ELECTRONIC WORKSHOP MANUAL AND GUIDE

Electronic Workshop Manual and Guide

CARL G. GROLLE

Parker Publishing Company, Inc.

West Nyack, New York

PARKER PUBLISHING COMPANY, INC.

West Nyack, N. Y.

Library of Congress Cataloging in Publication Data

Grolle, Carl G
Electronic workshop manual and guide.

Includes index.
1. Electronic apparatus and appliances--Maintenance and repair. I. Title.
TK7870.G763 621.381'028 76-49514
ISBN 0-13-252502-X

Printed in the United States of America

This book is dedicated in gratitude to my parents,
Carl and Dorothy Grolle

OTHER BOOKS BY THE AUTHOR:

Complete Guide to Electrical and Electronic Repairs. Parker Publishing Company, Inc., 1975.

Electronic Technician's Handbook of Time-Savers and Shortcuts. Parker Publishing Company, Inc., 1974.

How This Book Will Help You Save Time and Money in Electronic Servicing

Time is money. When an electronic technician can work faster or more efficiently, his profits will soar. This book is written to help you increase your productivity . . . and your profits.

How many problems in electronics could be solved quickly and easily if little problems weren't constantly bugging the technician. For example, a simple matter such as bench space can be a problem. Electronic apparatus, test equipment, hand tools, schematics, test leads, etc., can overwhelm a person if he is not systematically organized. Chapter 1 will show many practical ways a test bench can be improved to utilize the maximum space in the minimum area. Ideas for locating key test instruments and hand tools in the most accessible locations will be described.

Did you ever wish you could be on both sides of a television chassis at the same time? You can . . . well almost, by using a strategically located mirror system built into your bench. You'll wonder how you ever managed without it.

Every electronic technician knows that good service information means speedy repairs. The trick is locating the right repair information when you need it. Chapter 3 is packed with ideas on locating and storing service information. Remember all the times that you had the information you needed but couldn't remember where it was hiding? Chapter 3 outlines a filing system that can't be beat.

Tube and transistor checkers can be lifesavers, or they can really bog you down. How do you know if that tube you just checked is really good? Or how about the transistor's leads: which is the base, emitter, and collector? Read about tube or transistor checkers in Chapters 5 and 8, and how they can be more helpful to you.

Are electronic surplus parts any good? Can you make more profit using these parts? Is it worth the effort to seek out surplus parts that fit your needs? Millions of dollars in electronic parts are dumped on the surplus market annually. The man who can recognize and find useful parts is going to make money. Read about how the surplus parts field can help build your profits (Chapter 6).

Some of the least understood and most abused pieces of test equipment are the various types of probes. When should you use a low capacitance probe? Should a demodulator probe be used for checking the composite video waveform? What's the best kind of probe for checking 30,000 volts? Chapter 7 answers these probe questions and many more. Don't let test leads get you down. If you have ever done a TV sweep alignment you might have felt like you were in the middle of a test lead spaghetti patch. Read about tricks and kinks that straighten out test lead problems.

Some of the easiest and fastest to use test equipment can be built by you in a jiffy. The majority of electronic bugs can be found without elaborate test equipment. Build a low cost signal injector, a foolproof transistor-diode checker, a tester for SCR's and triacs, a unijunction analyzer, and laboratory power supply. These instruments are great! They work. They work fast, and they find troubles.

Chapter 11 is devoted entirely to checking solid state circuits. Everything is covered from silicon diodes to integrated circuits. Troubleshooting solid state is "duck soup" when you use the right techniques. Learn how voltage and resistance measurements can make meaningful circuit diagnosis possible.

The final chapter in the book explains much about one of the best test instruments of all—the oscilloscope. The device tells all if you can set it up and interpret the waveforms. Master this complex instrument and you'll be able to find troubles that you didn't know existed. Read about triggered sweep scopes and how to set them up. A detailed explanation of the composite video waveform will explain the sync pulses, blanking pulses, video, and color information seen on the scope tube.

The previous paragraphs have detailed a few of the items in this book, and you will find hundreds of other worthwhile, money and time saving suggestions. This book is practical, straightforward, and easy to understand. It will be fun to read, save you time, and help you make money.

Carl G. Grolle

Contents

ELECTRONIC WORKSHOP MANUAL AND GUIDE

1 Electronic Workbench Improvement Ideas

The technician often overlooks the electronic workbench as an effective troubleshooting and repair aid. Many electronics men are so wrapped up in complex theory, circuit and instrument analysis that they tend to disregard the various ways an efficient workbench layout can make each job run smoother.

This chapter will point out some workbench ideas that really work. Workbench design, efficient use of space, drawers, doors, and accessories will be discussed.

1-1. PRACTICAL SPACE AND TIME SAVING DESIGN

It seems that there is never enough workbench space. By the time the test instruments, handtools, schematics, and other service aids are set up, there's no place left to work. Just making the work area extra large is not the answer because most of the equipment is needed in a compact area. The trick is to have everything at your finger-tips but out of the way when not being used.

The ideal workbench area could be laid out in the manner of an airplane cockpit or a spaceship control module. Here the pilot can reach all the instruments quite easily without too much effort. Yet, there's no feeling of clutter because everything has its place.

Figure 1-1 shows a workbench design that has been used for all types of troubleshooting, repair, and electronic construction. Notice the U-shaped design that permits the technician access to three separate areas of the bench merely by turning his body. A swivel stool with casters works out well with this type of workbench.

Most of the bench top area is free and uncluttered. The shelves around the back perimeter of the bench hold all of the test equipment. Keep the most used instruments—VTVM, DVM, VOM, scope, and generators, or whatever—close to the center where they will be most available. Position your other instruments towards the ends of the shelves depending on the amount of use each will see.

The workbench shown has two shelf layers. More shelf layers can be added vertically if the need arises. Most new test instruments are much smaller in size than their ancestors. Solid-state technology—transistors and integrated circuits—has made it possible to pack oodles of electronics in a small enclosure. This is convenient when you're using a multitude of instruments at once.

Figure 1-1 has drawer space at the ends of the bench but not in the middle. Don't let drawers (see Section 1-2) interfere with your working area. Keep them as close to you as possible, but not in your immediate work area. Whenever you are using something stored in a drawer, you'll leave the drawer open. If the drawers are at the ends of the bench, even though they are left open, they will not interfere with your work.

At the bottom back of the workbench are storage shelves for seldom used materials. Generally there is a lot of space in this area that is wasted. Keep these shelves back far enough so that you can't touch them with your feet or knees when you're sitting down.

Nearly everyone could use more space. The U-shaped design is practical because it fits readily into the space available. Make this workbench fit the space that you have and you'll really enjoy the ease of working that this design provides.

1-2. USEFUL TIPS FOR DRAWERS AND DOORS

No electronic workshop is complete without plenty of storage areas. Build drawers and cabinets into your workbench (Section 1-1). Drawers are handy because they make tools and parts easily accessible. You can custom make drawers to utilize

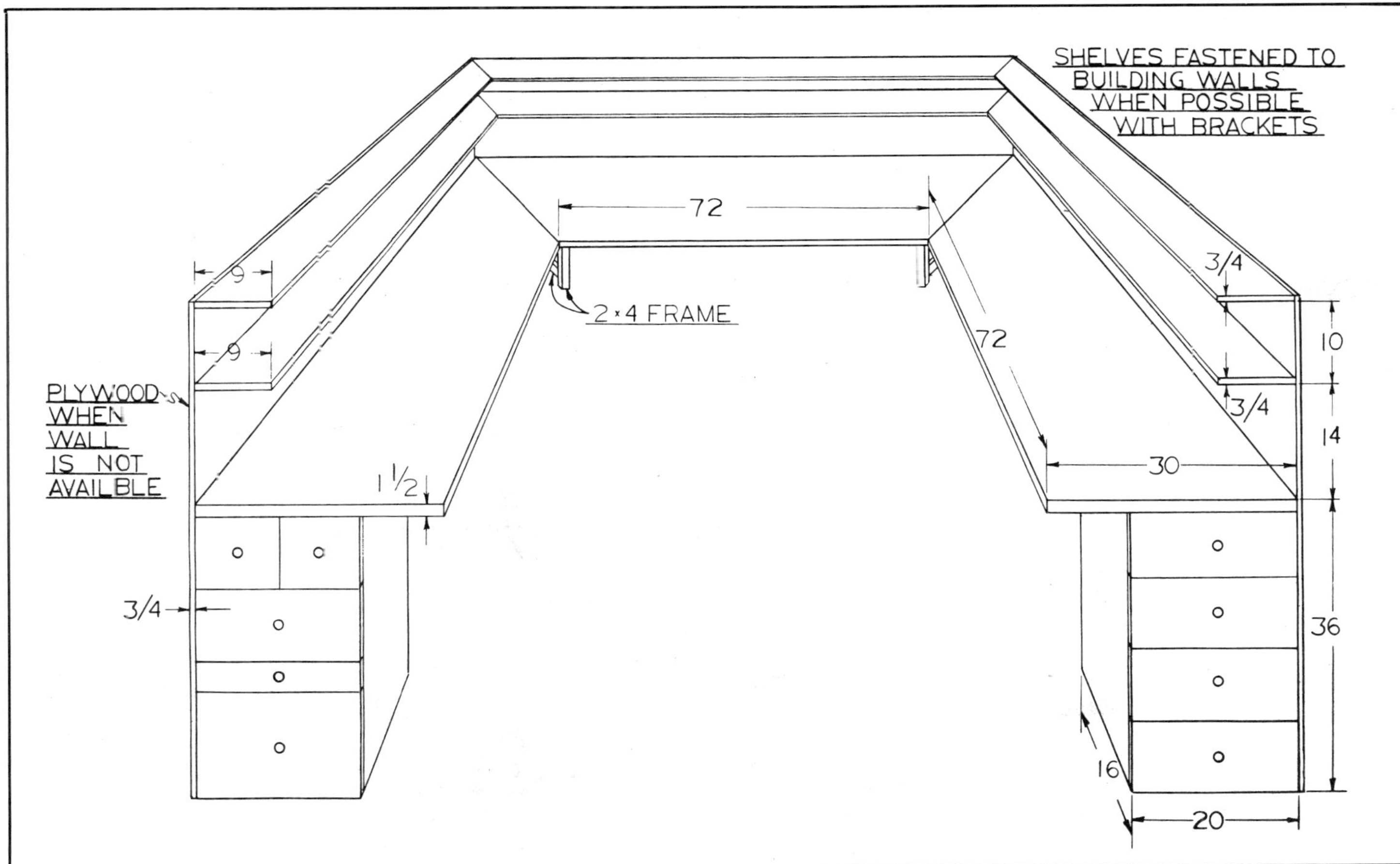

Figure 1-1. Space-saving workbench design.

the maximum space. Make your drawers various heights so no room goes to waste.

Drawers are easy to build and well worth the construction time. Figure 1-2A shows a simple drawer construction using plywood or flakeboard. Use these materials so that warping, shrinking, and swelling will be kept to a minimum. Don't try to fit the drawer into the frame with precision tolerances. Drawers work better if they don't fit too snug. Make the drawers slide easily by fitting drawer guides to the workbench frame. Rubbing these guides with soap or candle wax will make the drawers move effortlessly.

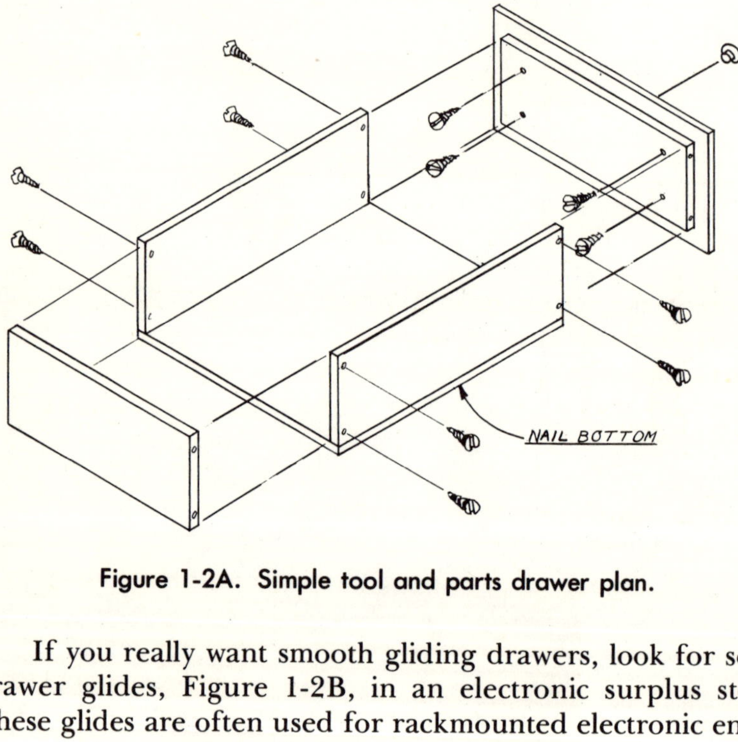

Figure 1-2A. Simple tool and parts drawer plan.

If you really want smooth gliding drawers, look for some drawer glides, Figure 1-2B, in an electronic surplus store. These glides are often used for rackmounted electronic enclosures so they can be pulled out for maintenance. These glides are expensive, but if you pick them up at surplus they are usually much cheaper. Fasten them to your workbench frame and connect to the sides of the drawers. Now when the drawer is pulled it will practically move itself!

Built-in file cabinets make good storage if you don't want

to build your own. Be sure to use full suspension drawers on easy-glide rollers. This type of drawer will pull out its entire length with minimum effort. The drawback to file cabinets for general storage is that the drawers are usually much larger than necessary for efficient use of space. Also, they are costly. Occasionally small drawer file cabinets will show up in a surplus store or in a garage sale. These units can't be beat for electronic storage.

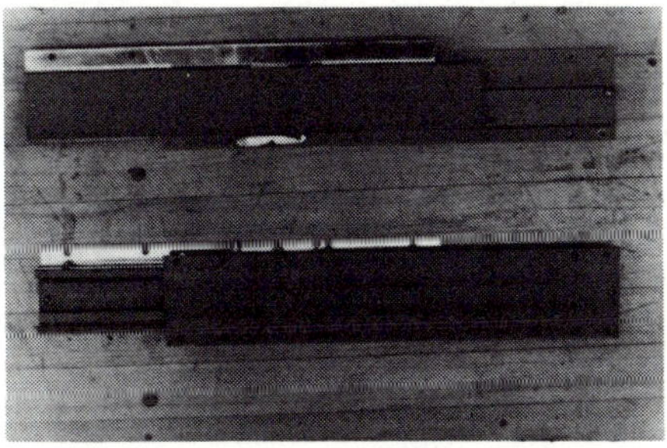

Figure 1-2B. Surplus electronic drawer glides.

Here's a fast, inexpensive way to obtain drawer storage. Use unfinished chests of drawers, which are available at most department stores, and build them into your workbench. Or utilize old chests of drawers, which are readily available at second-hand stores. Many of the older ones are constructed quite well and the drawers slide with ease.

You'll probably want to build one or more cabinets into your workbench as well as drawers. Figure 1-2C shows a fast,

Figure 1-2C. How to make cabinet doors.

simple method to make doors that fit out of 3/4″ plywood. This door design works well and does not require a skilled cabinet maker. The cabinet frame does not have to be perfectly square in order to make a good fit, and the hinges do not have to be recessed to either the door or frame. Magnetic or spring latches work equally well and are simple to install. Follow the directions on the package and you can't go wrong.

1-3. HOW TO SELECT BENCH TOPS

Keep in mind the following points when you decide on the material for your bench top.

1. Durable, tough, abrasive resistant
2. Availability
3. Ease of working
4. Surface texture
5. Insulating qualities
6. Cost

There are numerous materials that could serve as a workbench top. Hardwood, tempered hardboard, and laminated plastic are some of the common materials that make a good surface.

The bench top is going to get a lot of wear and tear. Most electronic chassis are made from sheet metal that will scratch and scrape. You'll want to be able to work at your bench without babying the top; yet the surface should stay relatively unmarred.

Strips of maple hardwood laid side by side, glued and bolted together probably produce one of the finest kinds of bench surface. If money is no object, this type of top will last a lifetime. Most technicians will settle for less than a solid maple top. One of the most durable yet inexpensive materials for a bench top is 1/4″ tempered hardboard. Sheets 4′ × 8′ are readily available in lumber yards. Large sheets are best since this eliminates almost all seams. This material laid over a solid surface will give excellent servicability. It also fits the six points previously mentioned for workbench material.

Another good material is laminated plastic glued on 3/4″

plywood. More expensive and colorful than hardboard, it provides a hard smooth surface. If you decide on this material you'll probably have to be a bit more careful since it will scratch under hard treatment.

If you choose hardboard or laminated plastic, finish the front edge of the top with a strip of wood, Figure 1-3. The front edge of the work surface really takes a beating. The wood edge trim will make the top look better and also give it more protection.

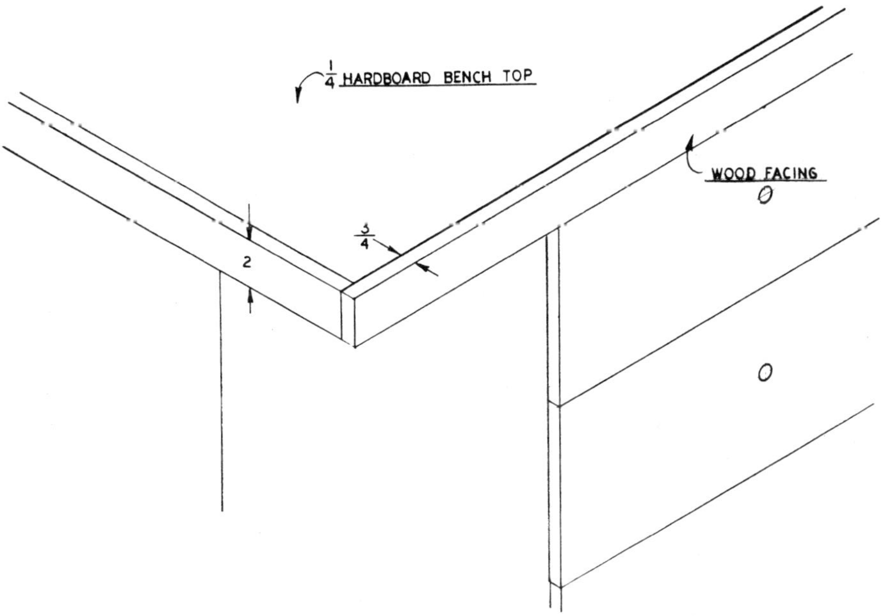

Figure 1-3. Wood trim for the workbench top front edge.

Regardless of the bench top material you choose, it's going to get scratched with normal working unless a few precautions are taken. Here's a tip using carpet that works like a charm. Obtain a few carpet remnants large enough to hold the equipment you repair. Place the carpet piece upside down on the bench surface. Set the TV chassis or whatever on top of the remnant. Now when the chassis has to be moved it will push easily in every direction. Your bench top will not become scratched and you will be able to work on all sides of the equipment effortlessly.

Another item that should be available for your bench top is a sheet of metal approximately 2′ × 4′. Sixteen-gauge copper or aluminum would be excellent. Normally it's not advisable to work on a conducting surface, but there are times when this is necessary. Occasionally an alignment procedure will require all instruments to be placed on a common grounded metal surface. When soldering and desoldering MOSFETs, manufacturers recommend that you and all tools should be grounded to prevent a static electric puncture of the MOSFET junction.

1-4. AUXILIARY DEVICES THAT TURN A WORKBENCH INTO A LABORATORY

When you repair television equipment it would be helpful to have eyes that can see around corners. Invariably when troubleshooting TV circuits, the picture tube is hidden from view. Since the cathode ray tube is so important for analyzing trouble clues, it is constantly being monitored by the repair technician. The usual solution is either to strain your neck trying to be in two places at the same time or to fight with a portable mirror that always seems to be in the wrong place.

Why not build a mirror into your workbench along the entire back length between the bench top and the first shelf? Long thin mirrors are readily available in discount houses. Several of them mounted end to end will do the trick. Once the mirrors are permanently installed you'll wonder how you ever managed without them. A simple mirror installation can do wonders easing the work load.

What about service information and schematics? Often there's no room left for these valuable repair papers. If you lay them next to you on the workbench top, they will soon be covered with tools and equipment and may become folded and mutilated. The trick is to have a special place at the bench just for service information. Once a place is reserved for these papers it's much easier to keep them clean.

There are several good methods to handle schematics and such. A cork board with thumb tacks above the bench area is very handy. If you have a wall that's close and accessible, mount the cork board on it. Often wall space is not available. If you have a high ceiling, mount the cork board directly over the bench on weighted pulleys. When the board is needed just pull it down. After you have finished, it can be pushed up out of the way.

Another good technique utilizes the sliding breadboard principle. Mount a section of plywood under the bench top so that it can be pulled out when needed. If you mask or tack information on this board, it will stay in place when it's being used or when it's out of sight.

Every electronic workbench should have a test panel, Figure 1-4A, full of jacks leading to various electronic equipment:

1. Antenna connections
   VHF TV
   UHF TV
   FM
   Automobilc radio
2. Variable AC power supply
3. Variable DC power supply
4. Isolation transformer
5. Audio amplifier
6. Audio output transformer
7. Test speakers

Four sets of antenna terminals are included for TV, FM, and auto mobile radio. A variable 10A, 0-117 VAC power supply monitored by voltage and current meters is a must. A metered 3A, 0-30 VDC power supply is also necessary. For safety and line isolation be sure to include a 1:1 500W isolation transformer. A separate audio amplifier is often needed in design and repair work, so include one in the test panel. High and low impedance amplifier input jacks are provided as well as the output jack. The primary and secondary terminals of an audio output transformer are brought out on banana jacks. Two test speakers are also included, a 3.2 ohm and a 16 ohm. It's wonderful to have all these key items just a "plug" away from being operational.

Figure 1-4B shows the schematic wiring diagram of the test panel. Use a wide band TV-FM antenna coupling the signals through a signal splitter to their respective panel terminals. An

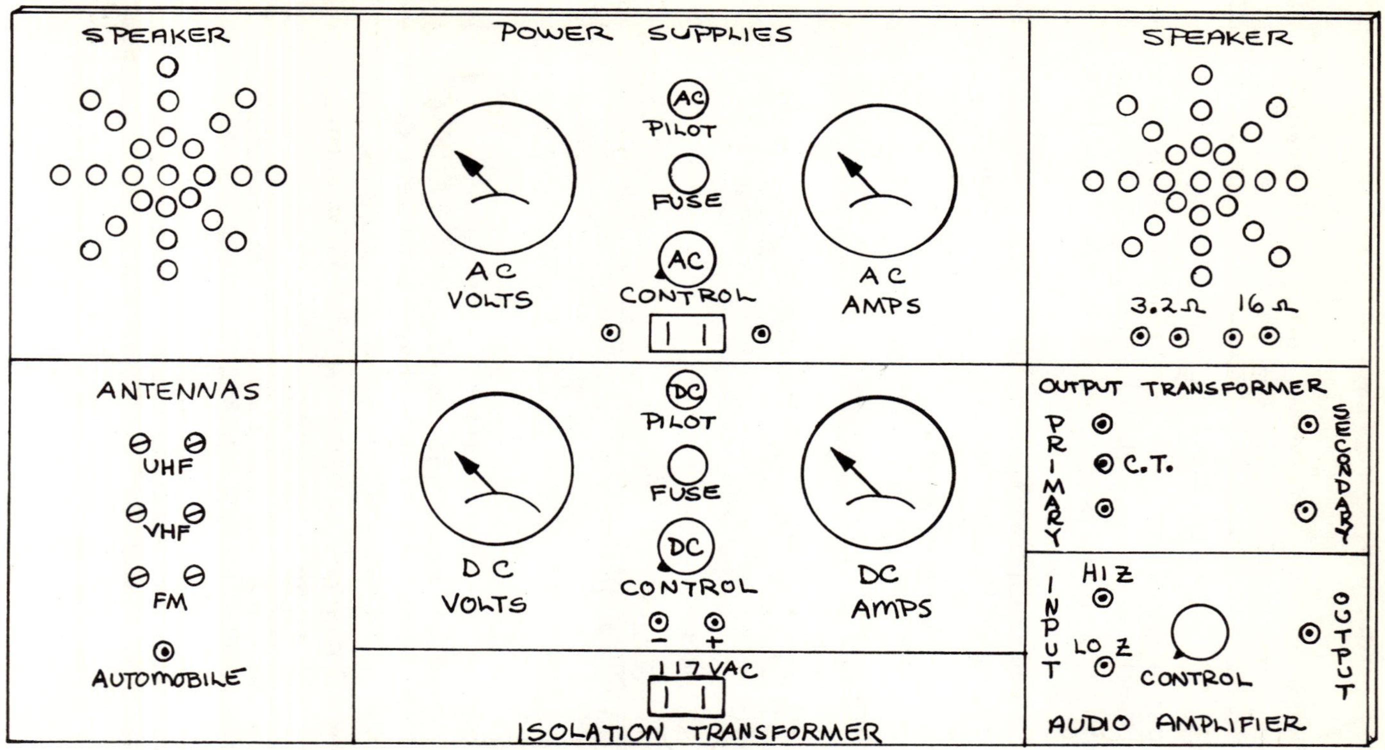

Figure 1-4A. Test panel front view.

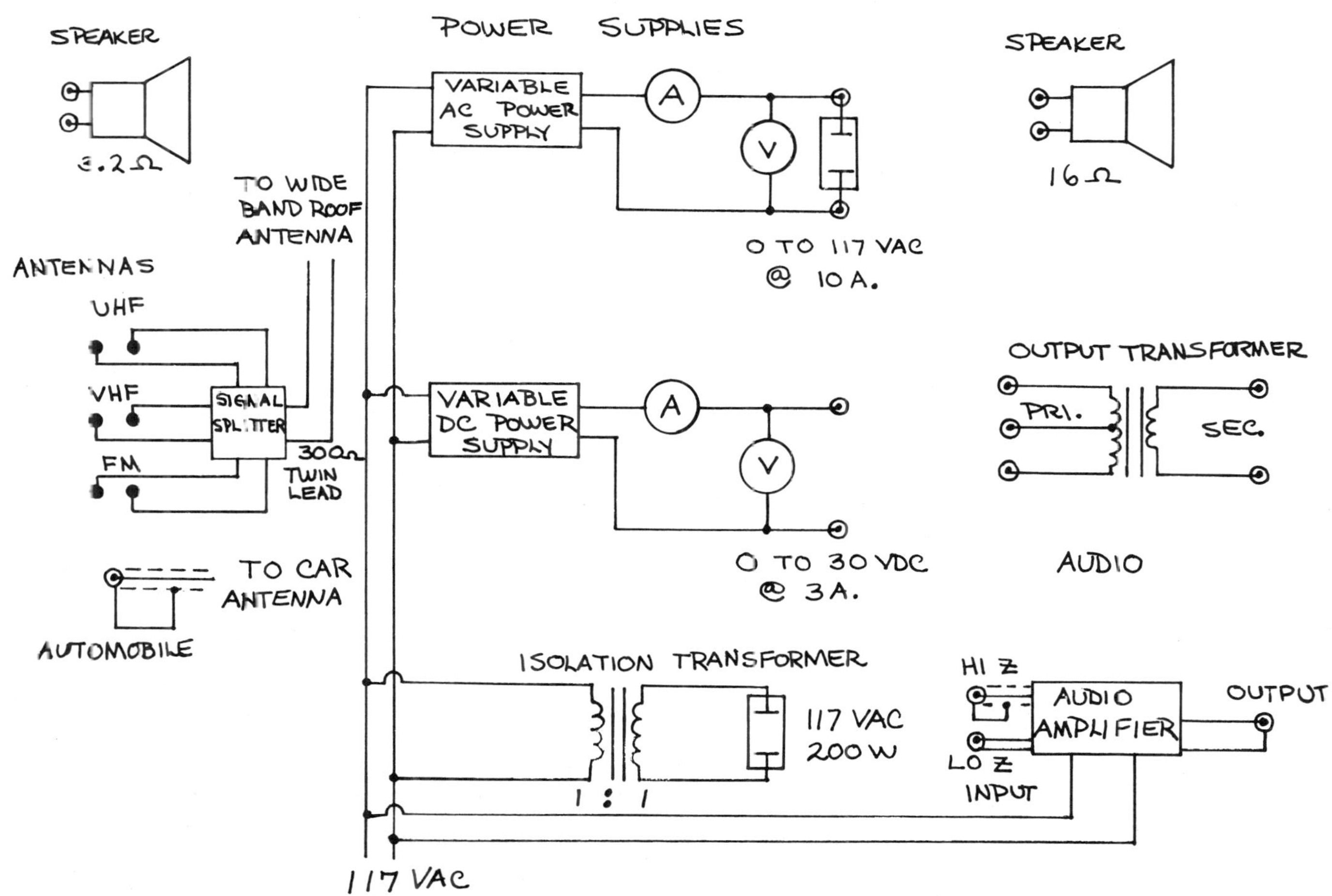

Figure 1-4B. Test panel schematic wiring diagram.

ordinary monopole telescoping car antenna is sufficient for the automobile radio antenna jack. Build the power supplies yourself, Sections 8-5 and 8-6, or buy them already made and install them behind the control panel. Low cost audio amplifiers are available in most electronic stores. Try to keep the amplifier's input jacks physically close to the amplifier. Shielded wire may be necessary for these inputs.

There are countless other electronic items that could be included in the test panel. The ones shown have proved extremely useful over the years for electronic repair and design work. Each technician will probably want to add a few pet ideas of his own.

1-5. TIPS FOR USING A WORKBENCH TEST PANEL

The test panel described in Section 1-4 can be used for multiple applications. However, before the test panel can become operational, a set of test leads should be made. Table 1-5 shows a selection of three foot leads.

Suppose there's a car radio lying on the workbench ready for repair. You're ready to "fire it up." Plug in the Motorola type plug to the radio's antenna jack. Connect the other end to the car antenna jack in the test panel. Clip black and red alligator clips to the radio speaker leads and plug the other ends of the clip leads into the banana jacks for the 16 ohm speaker. The only connections left are to the power supply. Set the DC power supply for 12 volts. Connect the negative power supply lead to the radio chassis (negative ground cars) and the positive lead to the "hot" power wire of the radio. That's it! The radio is operational and ready for dynamic troubleshooting.

The next electronic device to be repaired is a color TV. Set the TV on the workbench. Connect the appropriate test lead wires, Table 1-5, between the TV VHF and UHF terminals and the respective antenna terminals on the test panel. Plug one end of the cheater cord into the TV and the other end with the AC plug into the test panel isolation transformer receptacle for power and line isolation. Apply power to the TV, checking for picture, raster, and sound symptoms.

While we are discussing TV, here are some other ways the test panel can be of help. For some intermittent problems, re-

Quantity	End connector	Type of conductor	End connector
2	AC plug	Zip cord (line cord)	Two insulated alligator clips
2	AC plug	"Cheater cord"	Female receptacle
1	Motorola type plug	Coaxial antenna lead-in	Motorola type plug
2	Red banana plug	Test lead wire	Red banana plug
2	Black banana plug	Test lead wire	Black banana plug
3	Red banana plug	Test lead wire	Red insulated alligator clip
3	Black banana plug	Test lead wire	Black insulated alligator clip
2	Phone pin plug	Audio shielded cable	Red insulated alligator clip Black insulated alligator clip
2	Spade lugs	TV twin lead-in	TV lead-in clip

Table 1-5. Useful test panel leads.

ducing or increasing the line voltage may cause the problem to appear or stabilize. Simply plug the TV cheater cord into the variable AC power supply and try different line voltages. For some TV AGC troubles, and during IF alignment procedures, AGC action should be stopped. The usual method is to clamp the AGC line with the correct polarity of DC voltage. The test panel DC power supply jacks will provide this clamping voltage in a jiffy.

Here's a partial list of applications for the test panel audio amplifier:

1. Substitute audio amplifier for a defective record player
2. Audio signal tracer
3. Amplifier for an FM tuner that's being repaired
4. Audio substitute for a TV sound problem
5. Listening for low frequency pulses
6. Testing microphones and PA systems
7. Substitute for a problem audio section in a tape recorder
8. Test amplifier for design work
9. Aural indication of AF generator outputs
10. Intercom substitution amplifier

The more you use the test panel audio amplifier, the more uses you will find for it. It's a great tool!

Figure 1-5 shows the test lead arrangement for the test panel connected as a power microphone or small PA system. This illustrates the flexibility of the test panel.

Don't overlook the fact that the audio output transformer primary and secondary windings can be interchanged. If you need a step-down transformer, connect the input signal to the primary and the output will be a voltage step-down of the primary. However there are times when a step-up transformer is needed. Color organs and intercoms are two examples. The audio output transformer can still be used. Merely connect the input to the secondary, and the output will be voltage stepped up on the primary winding.

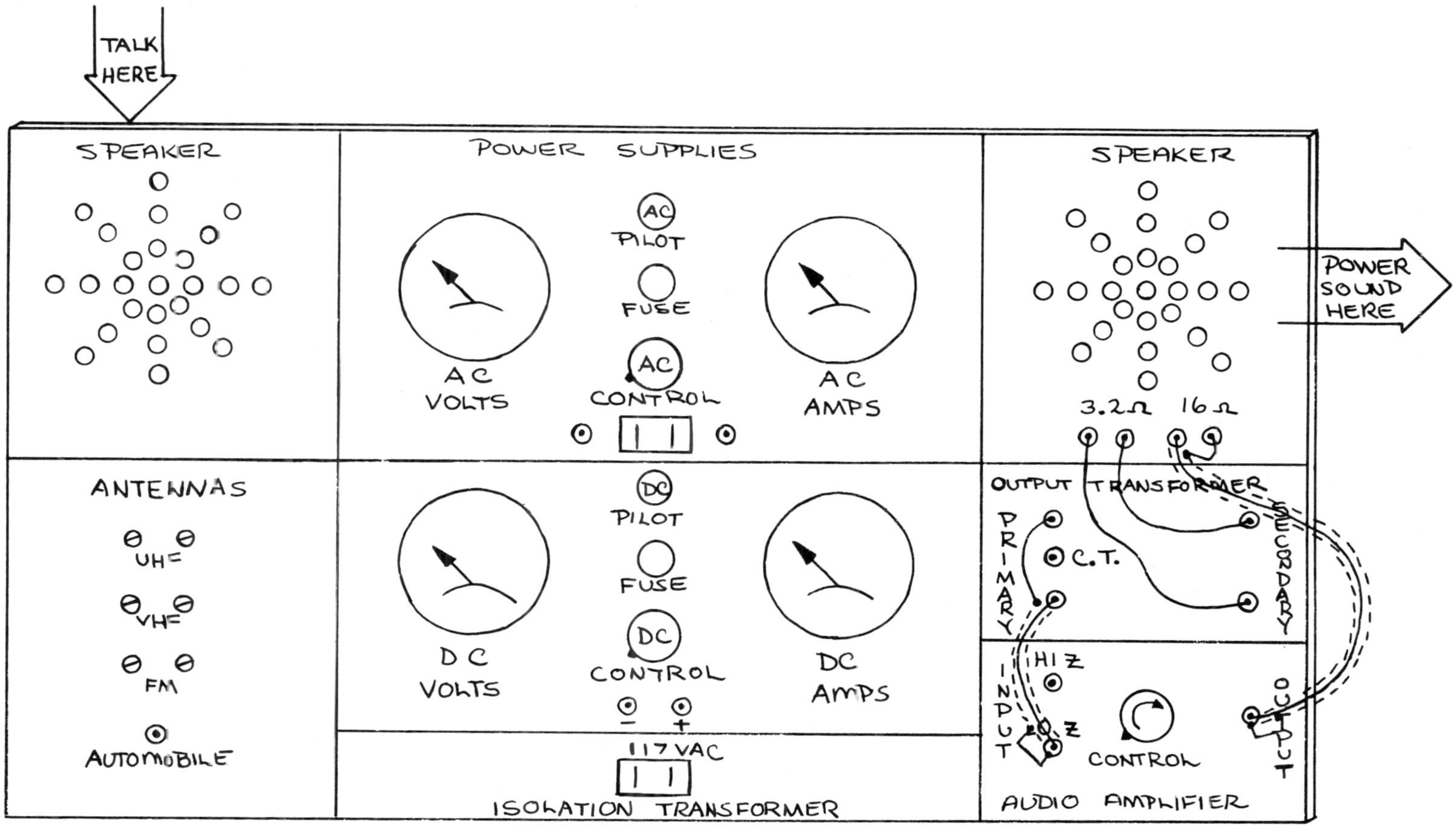

Figure 1-5. Test panel hooked up as a power microphone.

2 Easy-to-Use Ways for Storing Electronic Parts and Tools

In electronics, there seem to be a million parts needed at your finger tips. In addition to organization of these parts, and there are new ones being developed daily, the electronic technician must use more specialized tools.

This section will try to make sense out of the storage and organization of parts and tools. Useful space-saving ideas for storage cabinets will be discussed, along with some unusual ways of utilizing pegboard. Electronic hardware organization can be a problem. Find out some easy ways to categorize hardware. A surprising number of useful containers are routinely thrown in the rubbish. Take advantage of the serviceability left in these containers for your profit.

2-1. STORAGE CABINETS THAT REALLY WORK

One of the handiest cabinets for electronic parts storage is shown in Figure 2-1A. This commercial cabinet is really flexible in meeting your storage needs. The drawers can be easily removed for bench use, yet they will not pull out accidentally, spilling parts over the floor. There are three standard size drawers that will usually accommodate most parts. Each drawer can be partitioned with removable dividers. This gives you the option of using the entire drawer for one set of parts or dividing it into as many as ten separate compartments. There is recessed space for card identification on each drawer front and divider. Another fine feature of this cabinet is that it can be made secure by closing and locking the two front doors.

Figure 2-1A. Electronic parts storage cabinet with multiple sliding drawers.

Smaller cabinets that are also invaluable for parts storage are shown in Figure 2-1B. These cabinets are usually made of sheet metal with sliding plastic drawers. Don't skimp when you buy these units; buy the biggest one available. These drawers are similar to the ones described for the cabinet in Figure 2-1A, but they are smaller and more delicate. Don't try to pack too much in any single compartment or drawer because the parts will stick up and hook on the shelf above as the drawer is removed. When this happens you'll have to fiddle around trying to free the drawer and in the process a couple of drawer dividers will probably be pushed out of place—really making a mess. In the long run it's less costly in time and patience to partially fill the drawers and use more cabinets. When you buy these cabinets make certain that they can be stacked, saving you space and making a neat installation.

Figure 2-1B. Small parts cabinets.

Figure 2-1C shows a trim way to store wire. Note the top row of wire spools. It contains ten rolls of number 22 gauge solid hook-up wire. This wire is really a work horse for many electronic applications. The arrangement of spools is color-coded. The left hand spool is black followed by brown, red, orange, yellow, green, blue, violet, gray, and finally white. Slide the wire spools on to the center pole alternately, so that, for example, if the wire is coming off the top of the white spool, the gray spool will have the wire coming from the bottom. The next spool will have the wire coming off the top again, and so on. The reason is that when wire is pulled from individual or multiple spools, minimum interference will occur between adjacent spools. To keep the wire ends neat and in place, drill and grommet a hole in the shelf directly below each spool and feed the wire through.

The second shelf pole also contains wire spools. Even though these spools are various sizes, keep them installed in an alternate arrangement. This pole contains line cord wire, stranded hook-up wire, coaxial cable, twin lead, test lead wire, heavier gauge solid wire, and heater appliance cord. Although this wire arrangement uses an entire shelf area in a cabinet it is well worth the space because of the convenience and ease of access.

Figure 2-1C. Wire storage cabinet.

Refer to Figure 2-1D showing two versions of a cabinet storage area. The area on the left has the solder, flux, fuses, cans, containers, and whatever thrown haphazardly into it. You are lucky if you can see the first row and in no time you will forget what's buried in behind. Now look at the right hand version of this same area. The only difference is that a few sub-shelves and dividers, easily made from scrap sheet metal, have been installed. Now everything in the area can be plainly seen and reached. No more rummaging and guessing the contents of a shelf area. You will be surprised how a simple rearrangement of a shelf area can make your work easier and more pleasant.

2-2. HOW TO FULLY UTILIZE PEGBOARDS

Hardboard pegboards are probably the most common type of tool hanging system. The reason they are so popular is that they are relatively inexpensive, convenient, and they work. This section will point out ways to make pegboards even better.

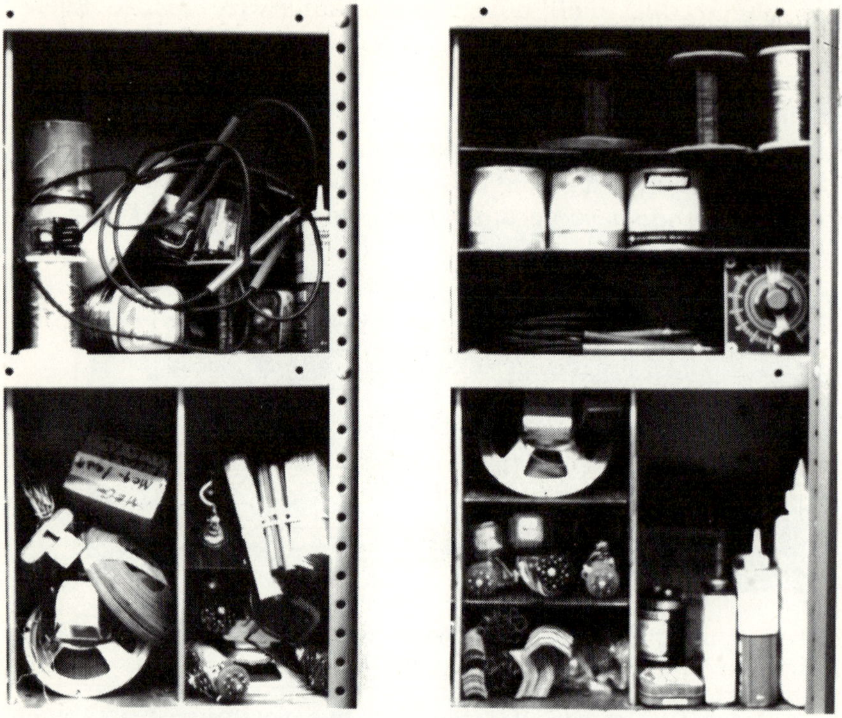

Figure 2-1D. Cabinet storage area—before and after.

One method to improve a pegboard tool installation is to mark the outline of the tool behind its hook. Painting the pegboard a light color such as tan or yellow and then outlining the various tool shapes with a felt tip marker is quick and easy. If you want to be a little more elaborate, fill in the outlines with various colors of paint. A color code could be worked out to fit the different categories of tools.

Cutting out the outline of tools in light cardboard and white-gluing the shapes on the pegboard is also a good way to mark tool positions. Commercial die-cut markers with pressure sensitive adhesive backing are available. They are inexpensive and easy to apply. Some of these markers can be removed and reused in new locations.

A problem that often arises when using pegboard hooks is the tendency for the hook to become loose when a tool is removed. The hook shown in Figure 2-2A eliminates this headache. Once this hook is "planted" it will stay put until you want it changed.

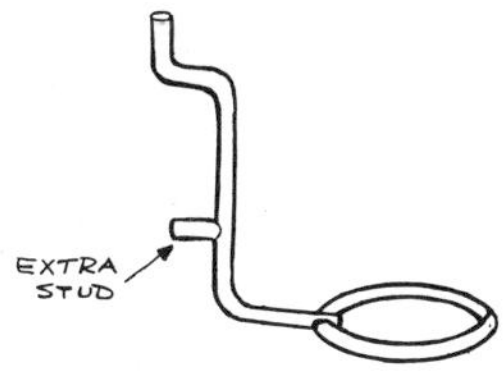

Figure 2-2A. Jiggle-proof pegboard hook.

There are all shapes and sizes of pegboard hooks. However invariably there will be a few tools that do not fit satisfactorily in any of the standard configurations. It's easy to make your own tool holders using existing hooks as the foundation. A little sheet metal, some ingenuity, and a few standard hooks can make all kinds of specialized brackets. Figure 2-2B shows a few of these custom made tool holders. Wire 1/8" to 5/32" in diameter makes good hooks for out-of-the-ordinary shapes. You'll probably have to experiment bending the wire but once you obtain a good design, these homemade hooks really do the job.

When talking about pegboards just about everyone thinks of a large perforated hardboard panel holding a multitude of tools. Don't forget pegboard tool holders can be small, too. Any place where there's a little space wasted is a spot for a mini-pegboard installation. These mini-pegboards are especially good for storing odd but related tools. For instance, where would be a better place for keeping jeweler's tools than a mini-pegboard? Make brackets for all the jeweler's screwdrivers, tweezers, miniature long nose and diagonal pliers, loops, and other tools of this nature. You'll probably find these tools are now scattered all over your workshop. Put them together and they will stay together. They won't take up much space. There's probably an ideal spot on the side of a bench or cabinet door that is accessible but unused. How about another mini-pegboard reserved for the newer tools needed for IC and printed circuit work? There are all sorts of applications for these little boards; get the mini-pegboard habit. You'll like it.

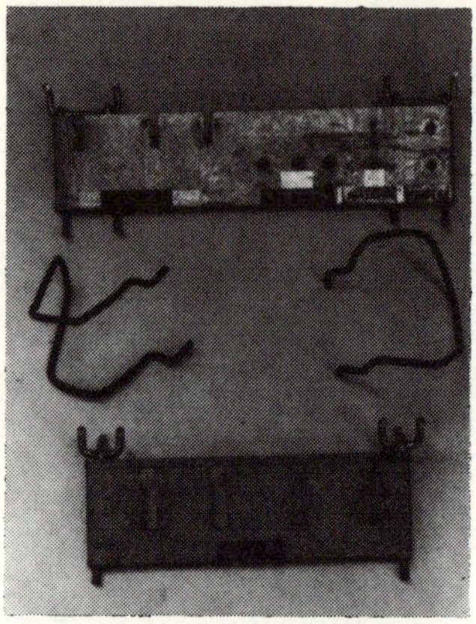

Figure 2-2B. Custom made pegboard tool holders.

When designing a large pegboard installation try to keep the following points in mind:

1. Utilize maximum space but don't overcrowd.
2. Keep tools of a kind together.
3. Position tools within your normal reach.
4. Don't block a tool removal path with another tool.
5. Use one hook for one tool.
6. Allocate some space for new additions.
7. Keep the most used tools in the easiest to reach area.
8. Plan for neat cord storage when hanging electric tools.

Planning a good pegboard set-up takes a fair amount of time and thought. However, in the long run an efficient installation will pay for itself a thousandfold.

2-3. EASY WAYS TO STORE ELECTRONIC HARDWARE

The secret to keeping some kind of order in electronic hardware is grouping of related items. Most electronic technicians don't need a separate storage area for every conceivable kind of hardware item. Nine times out of ten hardware in a drawer, bin, tray, compartment or rack that has related items stored with it will be sufficiently easy to locate.

Let's start with machine screws, nuts, and washers. Organization of this kind of hardware is not a hopeless task and actually can be done quite simply. The majority of screws and nuts used in the electronics field are gauge numbers 4, 6, 8, and 10. The gauge number refers to the thread diameters. Table 2-3A gives information on these common machine screw sizes. Once you have worked with these screws visual observation is all that is necessary to determine the various sizes.

Gauge #	Diameter (inches)	Threads per inch	
		Coarse	Fine
4	.112	40	48
6	.138	32	40
8	.164	32	36
10	.190	24	32

Table 2-3A. Common machine screws used in electronics.

Figure 2-3A shows a drawer containing #6 gauge machine screws, nuts, and washers; this drawer fits in the cabinet shown in Figure 2-1A. The drawer is partitioned by six drawer dividers filled as follows:

Drawer Sections	Contents
1st	#6-32 machine screw 3/8″ in length
2nd	#6-32 machine screw 1/2″ in length
3rd	#6-32 machine screw 3/4″ in length
4th	#6-32 hex head nut
5th	#6 flat washer
6th	#6 internal lock washer

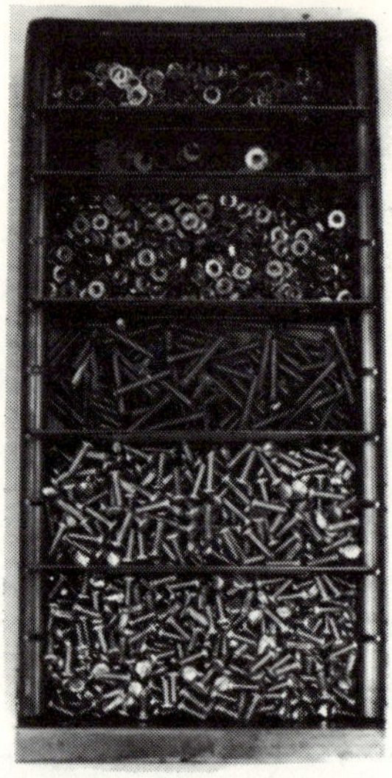

Figure 2-3A. Partitioned drawer containing #6 screws, nuts, and washers.

Any time #6 gauge hardware is needed just pull out the drawer and you've got a good selection. Arrange three other drawers in a similar manner containing #4, 8, and 10 gauge hardware respectively. If you want to stock fine threads per inch hardware, stock other drawers with them. A hardware selection similar to the one described will serve about 95% of your machine screw needs.

Of course machine screws, nuts, and washers can be separated into other kinds of categories. In addition to gauge number, threads per inch, and length, further machine screw separation can be made by type of head, material, head drive configuration, finish, and fit.

Another fastening hardware item that's used so often in electronics is the sheet metal screw. Organization of these screws can be divided into gauge number, screw length, type of head, and head drive configuration. As with machine screws, #4 and #6 gauge screws are the most common sizes for electronic applications.

The following list shows some of the useful categories in which electronic hardware can be classified:

1. Clamps
2. Clips
3. Fasteners
4. Female connectors (jacks and sockets)
5. Grommets, bumpers, and buttons
6. Handles
7. Insulators
8. Knobs
9. Lugs
10. Male connectors (plugs)
11. Nuts and washers for switches and pots
12. Spacers, posts, and rods
13. Terminal strips

Each one of these categories can be subdivided into as many sections as wanted.

See-through plastic storage boxes with multiple compartments make excellent devices to store electronic hardware, Figure 2-3B. Generally the largest box, about 10″ × 6″, with 18

Figure 2-3B. See-through plastic boxes for storing electronic hardware.

compartments is the most useful. These boxes are handy because you can see at a glance the contents without opening the lid. Four of these boxes will provide 72 separate compartments so it's easy to obtain a lot of organization in a relatively small area.

Figure 2-3C shows an inexpensive 40 bin storage rack that really does the job of separating larger electronic hardware items. A couple of these racks can do wonders organizing even the worst electronic hardware mess.

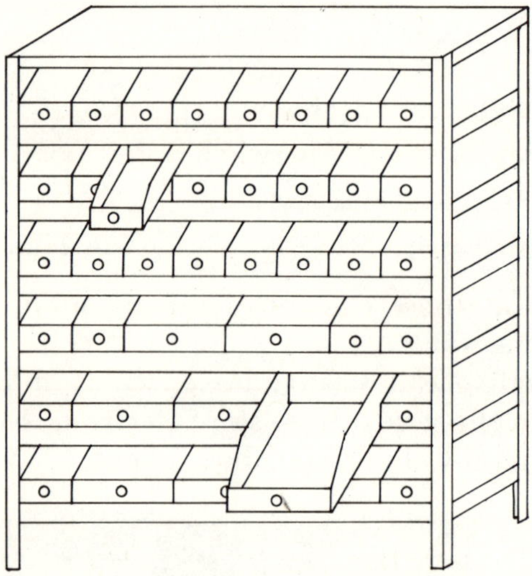

Figure 2-3C. Inexpensive 40-bin storage rack.

2-4. NOVEL TRICKS FOR THROW AWAY CONTAINERS

There are lots of ways to use containers that normally end up in the rubbish can. Figure 2-4A shows a rotating jar hardware holder made from empty olive jars and lids. The lids are fastened to the outside edge of a circular piece of 3/4″ plywood. The plywood is fastened to the ceiling or underside of a shelf through its center in a manner that allows it to rotate easily. Five jars contain machine screw hardware, one jar holds sheet metal screws, and the remaining jar is for assorted wood screws. It's a simple matter to spin the holder until the needed jar is closest, unscrew the jar from its lid and you're in business.

Figure 2-4A. Rotating jar hardware holder.

Baby food jars are another source of good containers. They make wonderful storage for small electronic hardware items. The lubricant, solvent, and cleaner holder, shown in Figure 2-4B, was designed with baby food jars. The jars contain:

1. White grease
2. Petroleum jelly
3. Bearing grease
4. Acetone
5. Alcohol
6. Cabinet polish
7. Liquid cleaner

These items are all needed from time to time in electronic servicing yet if kept in their original containers far too much space would be used. This little rack takes only a small amount of room but is convenient and accessible mounted close to your work area.

Figure 2-4B. Babyfood jar holder.

Need some divider compartments for a workbench drawer? Cut-down plastic gallon bleach bottles make excellent storage compartments. For small items cut off the top of a quart size plastic household ammonia bottle. They are usually round, but when squeezed together they will conform, wasting little space. Even cardboard milk cartons, pint, quart, one-half gallon, and even gallon size, make great drawer dividers. An organized drawer is going to save you a lot of rummaging and besides . . . the price is right.

While we're discussing bleach and ammonia bottles, the top half will make a tiptop funnel. The next time you transfer liquid from one container to another make a funnel and you won't spill a drop.

Every shop occasionally needs twine. How do you store that ball of twine? Cut a hole in the top side of a one-half gallon ammonia bottle and pop in the twine ball. Puncture a hole in the bottom for the string to exit and fasten the bottle to the ceiling by the handle. Now when you need some string merely reach up, and pull on the string end and it will unroll as pretty as you please. To make a super twine dispenser tie a used single edge razor blade on a string to the handle. It's convenient; saves you from hunting for a knife or scissors.

Think twice before throwing away pill bottles. They make ideal containers for small seldom used parts. Here's a list of a few items that you may want to store in pill bottles:

1. Ferrite cores for tuned transformers
2. Recessed hex head screws
3. Rivets
4. #0, 1, 2 screws and nuts
5. C rings
6. Miniature springs
7. PC drill bits
8. Switch contacts
9. Tiny soldering iron tips
10. Spray can valves

Foam plastic meat trays make a speedy breadboard base, Simply poke the parts leads through the tray bottom, following the schematic for layout position. Then connect the leads together with hook-up wire and solder to finish the circuit. Figure 2-4C shows an electronic siren breadboarded in this manner. After the circuit has been debugged and tested, tear down is quick and easy.

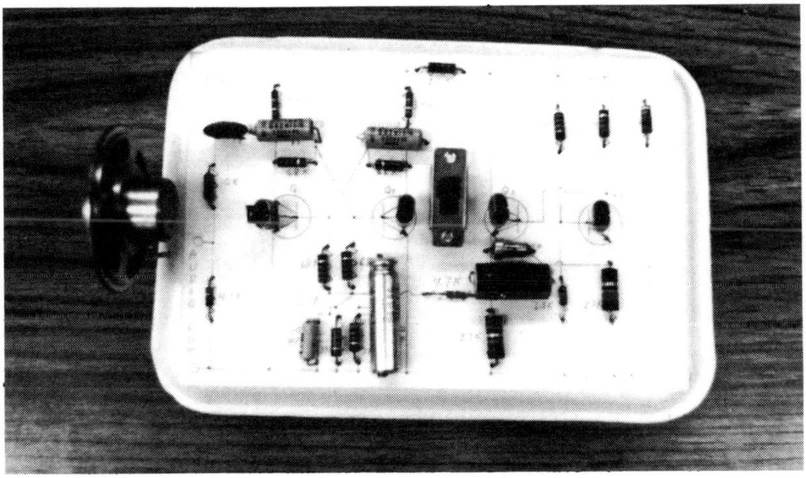

Figure 2-4C. Foam plastic meat tray used as a breadboard base.

Don't throw away those old chests of drawers; they make super, inexpensive parts cabinets. Sawing off the legs and squaring the top will save space and make a modern looking cabinet. A coat of paint and a little wax on the drawer slides is probably all that's needed to make the chest ready to go. A couple of these chests stacked will provide a surprising amount of storage space. If Aunt Sadie doesn't have a couple of these old chests stashed in her attic, you can pick them up at a garage sale or at a second hand store for next to nothing.

3 Time Saving Techniques for Locating and Using Service Information

One of the major drawbacks when repairing electronic equipment is obtaining schematics and repair information. This section will point out various methods that will help solve the service data dilemma.

Keep your service information where it can be retrieved easily. Read about a filing system that will allow you to put your hands on the right information instantly. Section 3-2 will point out various ways for utilizing the wealth of useful data found in electronic magazines.

Read about how the public library can help you. You might be surprised to find how much service data is available at libraries just for the asking. Learn about time-proven methods for obtaining the data you need from electronic manufacturers.

Section 3-5 will discuss how to keep an effective notebook to store all that invaluable, hard-earned trial and error information.

3-1. A FILING SYSTEM THAT MAKES MONEY

How many times have you needed a schematic that you knew you had but couldn't locate? Untold hours are wasted in electronic shops hunting for the service data that "disappeared." It's worth your time to organize a filing system for these important papers.

The first step is to obtain a cabinet(s) large enough to take care of your service information. Inventory your data because there is probably a lot more information floating around the shop than you realized. One central location for your papers is best. Then everything you need can be found and replaced with no unnecessary footwork.

A full-suspension four-drawer file cabinet is ideal. Don't skimp and buy a cheaper type because you'll regret it every time the files are used. The full-suspension drawers glide easily on nylon rollers whether the drawer is empty or jam-packed. Also they pull out the entire drawer length so even the very back corner is accessible.

Manilla file folders are a must. About 100 folders per drawer is adequate. These folders are inexpensive and tabbed in a systematic manner for easy access, Figure 3-1. Identifying each folder on its tab with a felt tip marker works well.

Figure 3-1. File drawer using tabbed manila file folders.

There are probably a zillion ways to organize material in a file cabinet. However, the system to be described is simple, straight forward, and has been used effectively for years in an electronic shop. Determine the main categories of data to be filed and reserve drawer space for each. Here is a list of categories you might want to consider:

1. Manufacturer's service information
2. Test equipment manuals
3. Schematics
4. Business records
5. Electronic projects
6. Catalogs
7. Training material
8. Electronic magazines

Of course the number of drawers devoted to each category is determined by your individual situation. Some categories will have to be broken down into further subdivisions. For instance, manufacturer's service information could be subdivided into television, radio, tape recorders, stereo, etc.

Once the main categories are determined it's just a matter of loading the manila folders with the appropriate material. Print the subject of each file folder on the index tab and keep in alphabetical order. Try to pick titles that really identify the folder contents or you will likely forget under what letter the material is filed. Alphabetical indexing works the best because it allows insertion of new folders any place in the system without disturbing the location of existing folders.

As soon as the file cabinets are set up, it takes only a few seconds to retrieve or store information. It's amazing how a little organizational effort can pay off in time saved and work simplified.

3-2. HELPFUL IDEAS UTILIZING ELECTRONIC MAGAZINES

Electronic magazines can be one of the least expensive and most worthwhile means that the technician can use to keep abreast of current innovations in the electronic industry. Magazine articles and advertisements are the only way many technicians can attempt to assimilate the flood of new developments that appear almost daily.

Keep your magazines well organized so information can be

retrieved easily when needed. A file drawer (Section 3-1) is an excellent place to store magazines. It's simple to pick out magazines when they are stacked in order with just the bound edge showing. The bound edge always has the title, volume number, and date printed on it.

Generally one magazine per year has the annual index printed in it showing alphabetically the magazine's contents for the previous year. It's a good idea to remove the annual index and file it separately. When you want to refer to an article, just pick out the annual index manila folder, scan the index for the article and in short order you'll have the volume and page. If you have ever tried to find an article by paging through old magazines, you'll appreciate the ease of this system.

A nice feature of many magazines is the reader service card, Figure 3-2A. This card has numbers on it that are keyed to advertisements in the magazine. When you want to learn more about a product circle the appropriate numbers, tear out the card and mail. It won't be long before the postman will deliver your package of information.

Courtesy of Howard W. Sams & Co., Inc.

AUGUST, 1975
OFFER EXPIRES NOVEMBER 1, 1975

Please indicate number of items you have circled→ ☐

1	2	3	4	5	6	7	8	9	10	11	12	13	14	15	16	17	18
19	20	21	22	23	24	25	26	27	28	29	30	31	32	33	34	35	36
37	38	39	40	41	42	43	44	45	46	47	48	49	50	51	52	53	54
55	56	57	58	59	60	61	62	63	64	65	66	67	68	69	70	71	72
73	74	75	76	77	78	79	80	81	82	83	84	85	86	87	88	89	90
91	92	93	94	95	96	97	98	99	100	101	102	103	104	105	106	107	108
109	110	111	112	113	114	115	116	117	118	119	120	121	122	123	124	125	126
127	128	129	130	131	132	133	134	135	136	137	138	139	140	141	142	143	144
145	146	147	148	149	150	151	152	153	154	155	156	157	158	159	160	161	162
163	164	165	166	167	168	169	170	171	172	173	174	175	176				

Print plainly or type below:

Name ....

Firm ....

Firm Address ....

City.................... State.................... Zip..........

Did you receive this issue through ☐ Distributor ☐ Subscription

Subscription Offer

Enter my subscription to ELECTRONIC SERVICING

☐ 3 Years $13.00 ☐ 2 Years $10.00 ☐ 1 Year $6.00

☐ Bill Me ☐ Remittance Enclosed ☐ Extend Present Subscription

IMPORTANT! Please check your business classification.

Type of Business

Check One

A. ☐ Independent Electronic Service Organization
B. ☐ Retailer with Electronic Service Department.
C. ☐ Independent or Self-Employed Service Technician
D. ☐ Electronics, Radio, TV Manufacturer
E. ☐ Industrial Electronic Service
F. ☐ Wholesale, Jobber, Distributor
G. ☐ Other (Specify) ....

Position

Check One

H. ☐ Owner, Manager
I. ☐ Ser. Manager
J. ☐ Technician
K. ☐ Other

Be sure you have checked one box in each column above!

Figure 3-2A. Reader service card.

The *Electronic Technician/Dealer* magazine has an especially helpful feature called "Tekfax." Manufacturers' circuit diagrams and technical information is published in the magazine each month, Figure 3-2B. The schematics are numbered and organized into groups with an index published periodically.

1587

ZENITH

TV Chassis 19FB14

ELECTRONIC TECHNICIAN/DEALER TEKFAX

MAY • 1975

COMPLETE MANUFACTURERS' CIRCUIT DIAGRAMS AND TECHNICAL INFORMATION FOR 5 NEW SETS

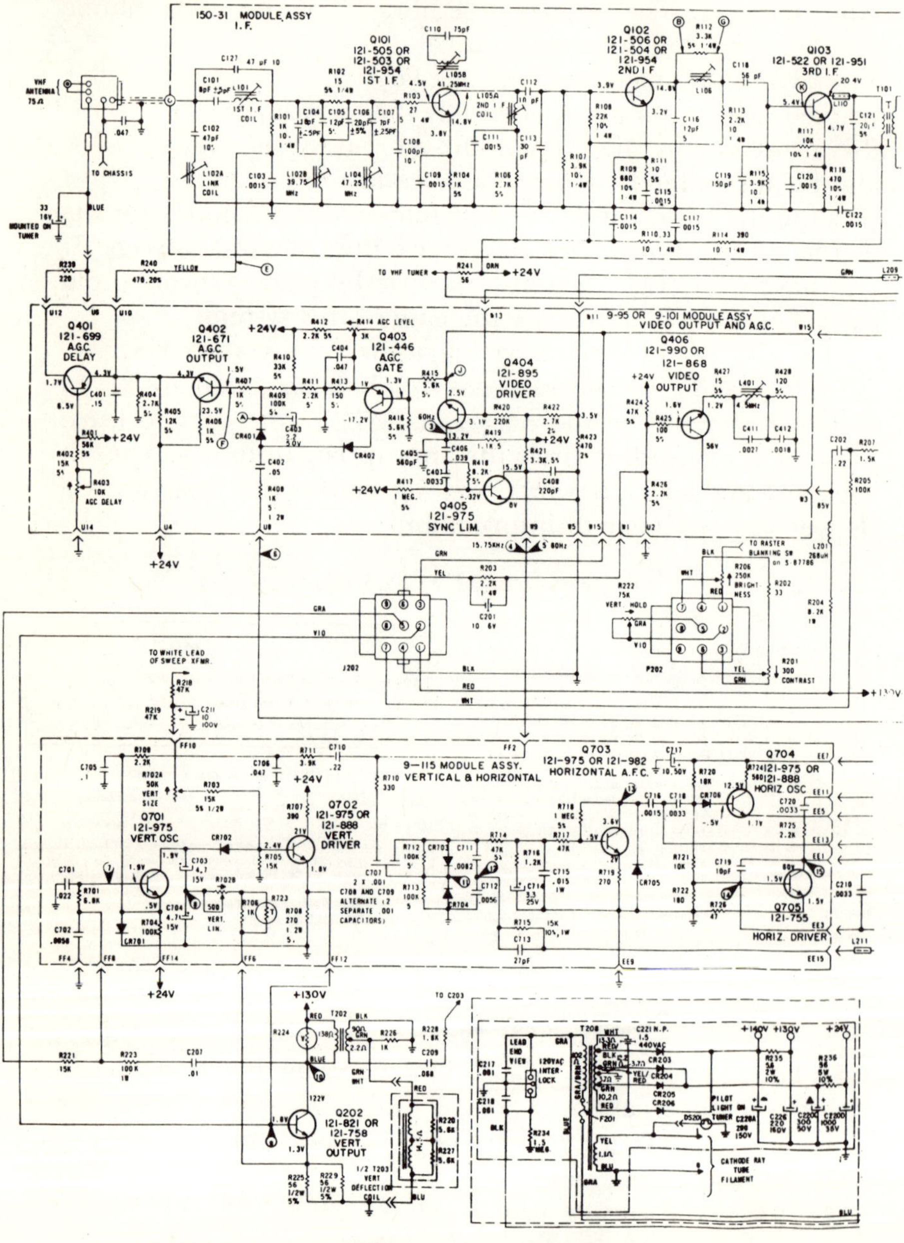

Figure 3-2B. Manufacturer's circuit diagram and technical information.

SYMBOL	DESCRIPTION	ZENITH PART NO.
C220A	200mfd electro capacitor 150v	
C220B	50mfd electro capacitor 150v	22-7314
C220C	300mfd electro capacitor 50v	
C220D	1000mfd electro capacitor 35v	
R215	varistor	63-5440
R224	varistor	63-10281
R414	3K AGC level control	63-10148
R702A	50K rotary dual rotary control	63-10225-01
R702B	500 ohm rotary	63-10225-01
R723	thermistor	63-10290
T201	audio output xformer	95-3120
T204	horiz driver xformer	95-2895-03
T205	sweep xformer	S-97079
T208	power xformer	95-3141-01
T1102	quad xformer	95-2620
F201	fuse .6a bel fuse	136-100

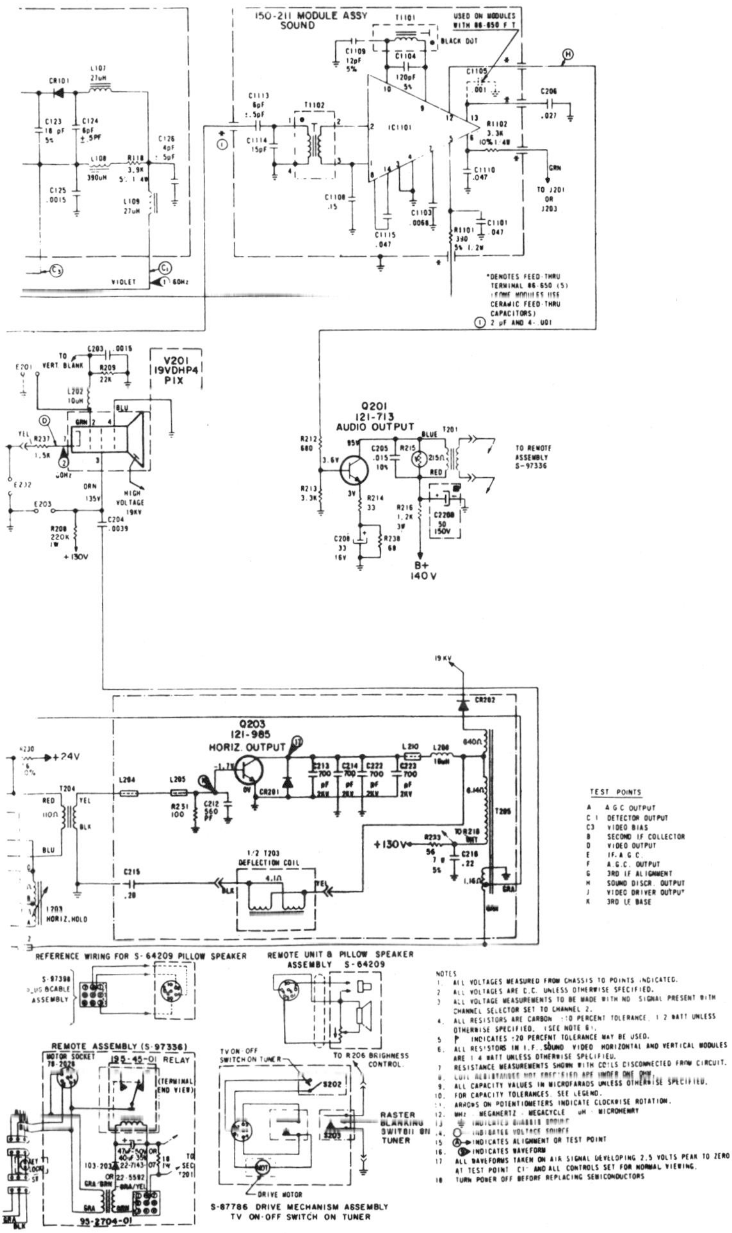

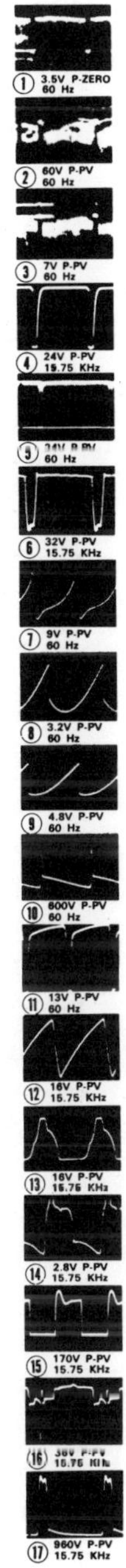

Courtesy of Electronic Technician/Dealer

The edges of the schematics are perforated and pull out of the magazine easily. Storing these service aids in a three-ring binder gives quick access. Once a quantity of these diagrams has been accumulated you'll find they represent a large proportion of the TV's that are in to be repaired.

These diagrams are a compact source of useful information. In addition to the schematic drawing, DC voltages are shown for all key areas. Important test points are clearly marked and identified. Even the description and part numbers for special components are provided. Waveforms identified with P-P V and frequency are included and keyed to the schematic diagram. By and large these diagrams provide all the information needed to repair the majority of TV defects.

If you are interested in construction of electronic projects, *Popular Electronics* magazine is one of the best. Their articles are well written, clear, and easy to understand. In fact many of the projects have been tested by PE before publication.

Many technicians subscribe to several electronic magazines in order to pick up as much new information as possible. Very often they reach you through the mail on or close to the same day. Don't make the mistake of trying to read all of the magazines in one evening. There's a strong tendency to do this since the articles will be of great interest to you. To properly read and understand a technical article, time and a good deal of thought are necessary. A good method to obtain the maximum benefit from a magazine is to leaf through it obtaining the "big picture" of the contents. While browsing, make a mental note of the articles that are of the most interest to you. Then, after you roughly know the contents, pick the most interesting article and read it thoroughly.

A common mistake when reading an electronic project or circuit description is to inadvertently skip analyzing the pictures, drawings, waveforms, or graphs that are being described in the article. Very often the "meat" of the article is represented in the illustrations and the words are just used for clarification. Make it a habit to always refer to the illustration when it's being described by the text. Give yourself enough time to thoroughly understand what the author is attempting to show. Speed reading may be alright for novels but when it comes to technical material it's a no-no.

3-3. PROFIT USING THE PUBLIC LIBRARY

You know that brick building that you drive by each day on the way to work? It's called a library; where other people go to sign out story books. All kidding aside, it's true that many technicians are so busy that they forget about the benefits a library can offer to them. All libraries do not have available the same amount or kinds of technical material, but most of them have more than you might think.

Sams Photofact® Folders, the stand-by electronic service information for countless electronic technicians, sometimes are stocked in the library, see Figure 3-3. For electronic home entertainment equipment service data the Photofact® Folders can't be beat. Any set manufactured from 1946 until the present is probably supplied in one of these folders. Over 93,000 models are covered—B & W and color TV, AM and FM radios, stereos, and record changers. The package includes the manufacturer's own PC board callout numbers, circuits, photos, schematics, replacement parts, and much more.

Use the Photofact® Annual Index to find the needed folder. Check the name of the equipment, usually the manufacturer's name, and the model number. Look it up alphabetically in the index. Opposite the model you'll see the Photofact® set and folder number. The library will have them stacked in order so it's a simple matter to pick out your folder.

Many libraries also stock one or more series of Sams specialized volumes:

AR — Auto Radio Series
CB — Citizens Band Radio Series
HTP — Home Tape Player Series
MHF — Modular Hi-Fi Series
SD — Scanner-Monitor Servicing Data
TR — Tape Recorder Series
TSM — Transistor Radio Series

Find the volume you need using the method previously described for locating Photofact® Folders.

Don't overlook the fact that many libraries stock a good selection of electronic related magazines. Back issues are usually obtainable as far in the past as five years. Here's a list of periodicals usually available in libraries:

1. CQ Radio Amateur's Journal
2. Electrical World
3. Electronics
4. High Fidelity
5. Popular Electronics
6. QST
7. Radio Electronics

Courtesy of Howard W. Sams & Co., Inc.

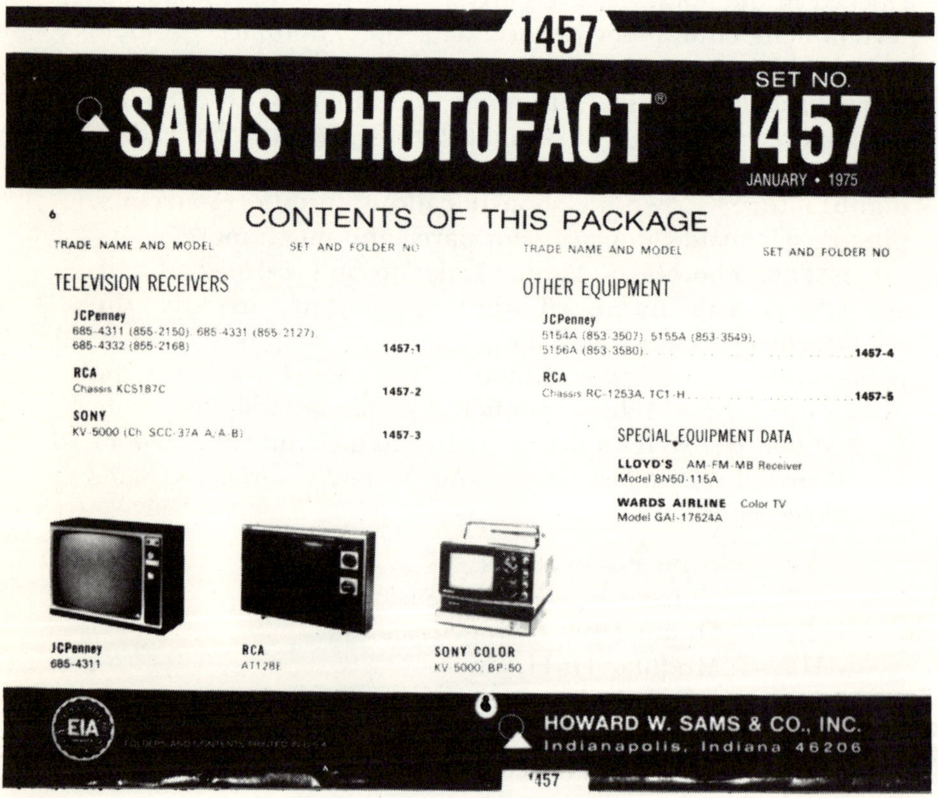
1457

SAMS PHOTOFACT® SET NO. 1457

JANUARY • 1975

6

CONTENTS OF THIS PACKAGE

TRADE NAME AND MODEL	SET AND FOLDER NO
TELEVISION RECEIVERS	
JCPenney 685-4311 (855-2150), 685-4331 (855-2127), 685-4332 (855-2168)	1457-1
RCA Chassis KCS187C	1457-2
SONY KV-5000 (Ch SCC-37A A/A-B)	1457-3
OTHER EQUIPMENT	
JCPenney 5154A (853-3507), 5155A (853-3549), 5156A (853-3580)	1457-4
RCA Chassis RC-1253A, TC1-H	1457-5

SPECIAL EQUIPMENT DATA

LLOYD'S AM-FM-MB Receiver Model 8N50-115A

WARDS AIRLINE Color TV Model GAI-17624A

JCPenney 685-4311

RCA AT128E

SONY COLOR KV-5000, BP-50

EIA

HOWARD W. SAMS & CO., INC.
Indianapolis, Indiana 46206

1457

Figure 3-3. Sams Photofact® Folder.

There are two periodical indexes, both arranged alphabetically, to use when referring to magazine back articles. *The*

Reader's Guide To Periodical Literature and *Applied Science & Technology Index* are the keys to unlock the secrets stored in past issues.

If you are an old time radio buff probably you will find service information tucked away on some dusty shelf. Want to know what type of output tubes are used in a Model 70 Atwater Kent Radio? Chances are you'll find that information and much more in one of the early editions of the Rider Manuals. These manuals give excellent service data on thousands of radio and television receivers. M. N. Beitman has authored numerous fine radio and television diagram and service information books that also cover a multitude of receivers.

Looking for an electronics dictionary or a one-volume electronics encyclopedia? You'll find radio-TV-electronic dictionaries as well as audio-electrical engineering-electronics encyclopedias. Numerous engineering and electrical handbooks are just waiting to be studied.

Other popular library books are those that prepare you for obtaining one of the many FCC licenses. Many of these books use a question and answer format similar to the questions on the actual test.

Need information about any phase of electronics? Chances are the library has the book(s) you need. Don't worry if libraries seem overwhelming; librarians are more than helpful and will go out of their way to help solve your problems.

3-4. ELECTRONIC MANUFACTURERS CAN BE HELPFUL

Some of the best service information, data, and the latest most up-to-date books and pamphlets can be obtained from electronic equipment and parts manufacturers. This information is available to the public at a very nominal cost. Read on to find the kind of literature that's available only a postcard away.

Training manuals, Figure 3-4A, are prepared by many of the major TV and electronic manufacturers. The manuals are generally very well written and illustrated, clear, concise, and practical. They are more than likely designed with the teach yourself concept in mind. Some use programmed learning and home study techniques.

Courtesy of GTE Sylvania

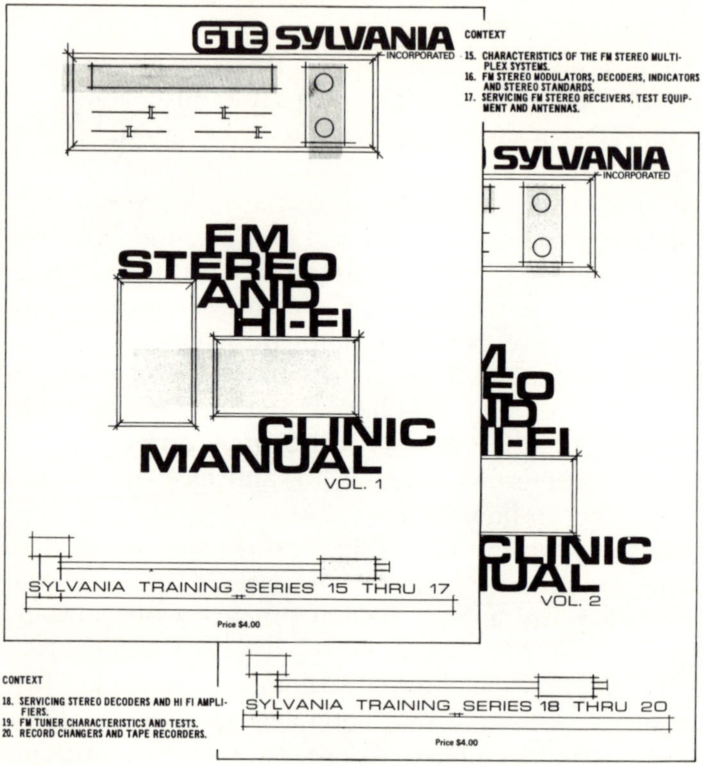

Figure 3-4A. Training manuals.

Service data, Figure 3-4B, for individual electronic apparatus can also be obtained from the manufacturer. Usually the data is very complete and may include the following kinds of information.

1. Electrical specifications
2. Installation and adjustment instructions
3. Instrument disassembly
4. Alignment procedures
5. Parts descriptions
6. Voltage waveforms
7. Resistance chart

8. Exploded views
9. Schematic diagrams
10. External wiring
11. Chassis layout
12. Printed circuit phantom view
13. Component location guide
14. Replacement parts

No shop should be without a semiconductor cross-reference guide, refer to Figure 3-4C. These invaluable booklets categorize the thousands of semiconductors and show how they can be replaced with a relatively small number of selected types.

Cross reference for low-power transistors, high-power transistors, rectifiers, zener diodes, field effect transistors, IC's, LED's, and others are usually indexed. In addition to cross-reference data these booklets usually show parameters and performance characteristics plus outline drawings detailing case and basing information. After you have tried to wade through the mass of semiconductors on the market these guides are worth their weight in gold.

Semiconductor specifications and application sheets are also available from the manufacturers. These sheets will tell you just about all you want to know about a particular semiconductor device. Figure 3-4D shows a typical specification sheet. The manufacturer outlines the following information for the metal oxide varistor.

1. Descriptive
2. Features
3. Benefits
4. Application
5. Graphs
6. Circuits
7. Ratings
8. Electrical characteristic
9. Dimensions

Courtesy of RCA Consumer Electronics

RCA

FILE
1975
C — 3

Television Service Data

Chassis
CTC 68 Series

GU 754

RCA Corporation
Consumer Electronics
Technical Publications
600 N Sherman Dr | Indianapolis, Indiana 46201

Model and Chassis Cross Reference

MODEL	CHASSIS	TMA	TUNERS	PICTURE TUBE
FU 505W*	CTC 68UD	441D	KRK 205JF, KF, LF/KRK 231LD, JD	21VBEP22
FU 510SEN*	CTC 68UD	441D	KRK 205JF, KF, LF/KRK 231LD, JD	21VBEP22
FU 518W*	CTC 68UJ	458J	KRK 205JE, KE, LE/KRK 231JC, KC, MC	25VCZP22
GU 550W	CTC 68UB	448B	KRK 199JC, KC, LC/KRK 231LJ, JJ	21VBEP22
GU 554L	CTC 68UB	448B	KRK 199JC, KC, LC/KRK 231LJ, JJ	21VBEP22
GU 558S	CTC 68UB	448B	KRK 199JC, KC, LC/KRK 231LJ, JJ	21VBEP22
GU 702W	CTC 68UE	458F	KRK 199JC, KC, LC/KRK 231JC, KC, MC	25VCZP22
GU 704L	CTC 68UE	458F	KRK 199JC, KC, LC/KRK 231JC, KC, MC	25VCZP22
GU 708S	CTC 68UE	458F	KRK 199JC, KC, LC/KRK 231JC, KC, MC	25VCZP22
GU 712S	CTC 68UE	458F	KRK 199JC, KC, LC/KRK 231JC, KC, MC	25VCZP22
GU 714L	CTC 68UE	458F	KRK 199JC, KC, LC/KRK 231JC, KC, MC	25VCZP22
GU 718S	CTC 68UE	458F	KRK 199JC, KC, LC/KRK 231JC, KC, MC	25VCZP22
GU 720W*	CTC 68UJ	458J	KRK 205JE, KE, LE/KRK 231KC, MC	25VCZP22
GU 724L*	CTC 68UJ	458J	KRK 205JE, KE, LE/KRK 231KC, MC	25VCZP22
GU 725H*	CTC 68UJ	458J	KRK 205JE, KE, LE/KRK 231KC, MC	25VCZP22
GU 728D, S*	CTC 68UJ	458J	KRK 205JE, KE, LE/KRK 231KC, MC	25VCZP22
GU 750W*	CTC 68UJ	458J	KRK 205JE, KE, LE/KRK 231KC, MC	25VCZP22
GU 754L*	CTC 68UJ	458J	KRK 205JE, KE, LE/KRK 231KC, MC	25VCZP22
GU 758D, S*	CTC 68UJ	458J	KRK 205JE, KE, LE/KRK 231KC, MC	25VCZP22

*ACM Models

CAUTION:

Before servicing this chassis, it is important that the service technician read the "Safety Precautions" and "Product Safety Notices" in this Service Data.

Service Data Reference

Service Data Contents — Page 3

First Edition — First Printing

CTC 68 Series

Figure 3-4B. Color television service data.

1975 C-3

SAFETY PRECAUTIONS

NOTICE: Comply with all caution and safety related notes located on or inside the receiver cabinet on the receiver chassis or picture tube.

1. When service is required, observe the original or correct lead dress. Extra precaution should be given to assure correct lead dress in the high voltage circuit area. Where a short circuit has occurred, those components that indicate evidence of overheating should be replaced. Always use the manufacturer's recommended replacement components.
2. Do not remove, install or handle the picture tube in any manner unless shatter-proof goggles are worn. People not so equipped should be kept away while picture tubes are handled. Keep picture tube away from the body while handling.
3. Protective shields are provided on this chassis for the protection of both the service technician and the customer. Protective shields removed for servicing convenience must be correctly re-installed and ANY MISSING SHIELDS MUST BE REPLACED. DO NOT OPERATE THIS INSTRUMENT WITHOUT THE PROTECTIVE SHIELDS IN POSITION AND PROPERLY SECURED.
4. When replacing a chassis in the cabinet, always be certain that all the protective devices are put back in place, such as: non-metallic control knobs, insulating fishpaper, adjustment and compartment covers/shields, isolation resistor-capacitor networks, etc.

   Before returning any instrument to the customer, the service technician must be sure that no protective device built into the instrument by the manufacturer has become defective or inadvertently defeated during service. Therefore, the following checks are recommended for the continued protection of the customer and service technician.

LEAKAGE CURRENT COLD CHECK

With the AC plug removed from the 120v AC source, place a jumper across the two plug prongs. Turn the instrument AC switch on. Using an ohmmeter, connect one lead to the jumpered AC plug and touch the other lead to each exposed metal part (antennas, handle bracket, metal cabinet, screwheads, metal overlays, control shafts, etc.), particularly any exposed metal part having a return path to the chassis. Exposed metal parts having a return path to the chassis should have a minium resistance reading of 2.1 megohms and a maximum resistance reading of 5 megohms. Any resistance value below or above this range indicates an abnormality which requires corrective action. Exposed metal parts not having a return path to the chassis will indicate an open circuit.

LEAKAGE CURRENT HOT CHECK

Plug the AC line cord directly into a 120v AC outlet (do not use an isolation transformer for this check). Use a Leakage Current Tester (RCA WT-540A) or a metering system which complies with American National Standards Institute (ANSI C101.1-1971 "Leakage Current for Appliances"), and Underwriters Laboratories (UL) 492, (ANSI C33.55). Measure for current from all exposed metal parts of the cabinet (antennas, handle bracket, metal cabinet, screwheads, metal overlays, control shafts, etc), particularly any exposed metal part having a return path to the chassis, to a known earth ground (water-pipe, conduit, etc). Any current measured must not exceed .5 milliamp. Reverse plug in the AC outlet and repeat test. ANY MEASUREMENTS NOT WITHIN THE LIMITS OUTLINED ABOVE ARE INDICATIVE OF A POTENTIAL SHOCK HAZARD AND CORRECTIVE ACTON MUST BE TAKEN BEFORE RETURNING THE INSTRUMENT TO THE CUSTOMER.

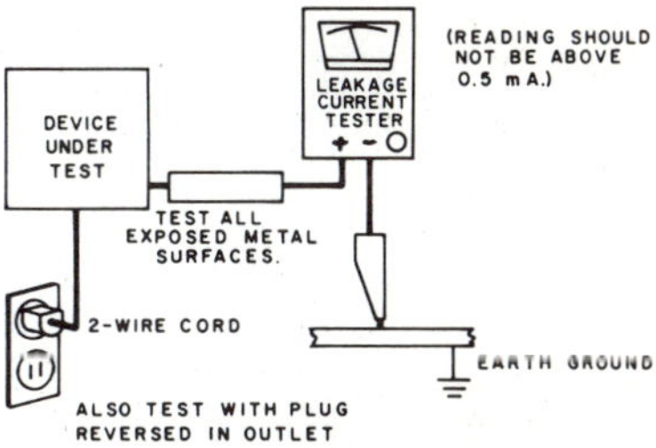

Figure 1. AC Leakage Test

X-RADIATION AND HIGH VOLTAGE LIMITS

The primary source of x-radiation in solid state television receivers is the picture tube. The picture tube is specially constructed to limit x-radiation emissions. The shields and mounting hardware for picture tubes have an x-radiation protection function and must be properly in place.

For continued x-radiation protection, the replacement tube must be the same type as the original, including suffix letter. High voltage is maintained within specified limits by the use of close tolerance safety related components/adjustments in the high voltage circuit. Refer to the TECHNICIAN X-RADIATION WARNING note on the Chassis Schematic in this Service Data for specific high voltage limits of this chassis. If high voltage exceeds specified limits, check these components specified on the chassis schematic diagram and take necessary corrective action.

PRODUCT SAFETY NOTICE

Many electrical and mechanical parts in RCA television sets have special safety-related characteristics. These characteristics are often not evident from visual inspection nor can the protection afforded by them necessarily be obtained by using replacement components rated for higher voltage, wattage, etc. Replacement parts which have these special safety characteristics are identified in this Service Data and its Supplements and Bulletins; electrical components having such features are identified by shading on the schematics and by (★) on the parts list in this Service Data and its Supplements and Bulletins. The use of a substitute replacement which does not have the same safety characteristics as the recommended replacement part, shown in the parts list in this Service Data and its Supplements and Bulletins, may create shock, fire, or other hazards.

Product Safety is continuously under review and new instructions are issued from time to time. For the latest information always consult the current RCA Service Data, Supplements and Bulletins. A subscription to, or additional copies of, RCA Service Data may be obtained at a nominal charge from your RCA Consumer Electronics Distributor or from RCA Technical Publications, 600 North Sherman Drive, Indianapolis, Indiana 46201.

Courtesy of General Electric Company

Figure 3-4C. Semiconductor cross-reference guide.

GENERAL ELECTRIC REPLACEMENT SEMICONDUCTORS 1975

GE celebrates fifteen years as a pioneer in the replacement market for Universal Transistors. This new Entertainment Semiconductor Replacement Guide evidences reaffirmation of the objective to continue adding universal types to the line only as they are needed to service the market.

The increasing number of solid state components used in all types of equipment signals matching growth in the number of types to be replaced. Selective additions since the last edition of this Replacement Guide reflect General Electric commitment to help both electronic distributors and servicing technicians balance the need to meet consumer requirements immediately yet maintain minimum inventory levels.

This 1975 edition of the Entertainment Semiconductor Replacement Guide contains 65% more cross-references than the previous edition, and will assist you in making needed repairs on both domestic and Far East equipment. Semiconductor products are listed in alpha-numeric order for ease in finding replacement numbers.

Courtesy of General Electric Company

ELECTRONIC INNOVATIONS IN ACTION

180.59 9/73
SUPERSEDES 180.59 12/71

GE-MOV™
Metal Oxide Varistors
95-1000 Volts RMS AC 130-675 Volts DC

GE-MOV™ VARISTOR SERIES

Description:

General Electric metal oxide varistors are voltage dependent, symmetrical resistors which perform in a manner similar to back-to-back zener diodes in circuit protective functions and offer advantages in performance and economics. When exposed to high energy voltage transients, the varistor impedance changes from a very high standby value to a very low conducting value thus clamping the transient voltage to a safe level. The dangerous energy of the incoming high voltage pulse is absorbed by the GE-MOV varistor, thus protecting your voltage sensitive circuit components.

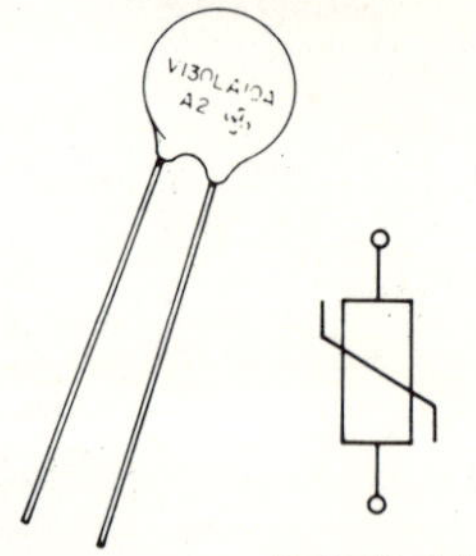

Electrical Symbol

Features:*

- Excellent Clamping (as low as 1.8 @ 10 Amps)
- Wide Range of Voltage Ratings (95 - 1000 Volts RMS) (130 - 675 Volts DC)
- Discharge Current Capability as high as 2000 Amps
- Energy Dissipation Up To 160 Watt-Seconds (Joules)
- Low Standby Drain
- Wide Operating Temperature Range
- Low Temperature Coefficient
- Fast Response (< 50 Nanoseconds)

*Refer to rating table for individual product capabilities.

Benefits:

- Promotes System Cost Reduction
- Improves Circuit, Component and System Reliability
- Increases Product Safety
- Eliminates Follow-On Current
- Reduces System Size and Weight Requirements
- Extends Contact Life
- Protects Circuit Insulation

Applications:

- Component and System Voltage Transient Protection in:
  - Inductive Switching Circuits
  - Transformer Switching Circuits
  - Regenerative Loads
  - Switching Magnetic Amplifiers
  - Semiconductor Reverse Recovery (Voltage)
- Contact Arc Suppression
- Reduction of Lighting Effects

Model Number Nomenclature:

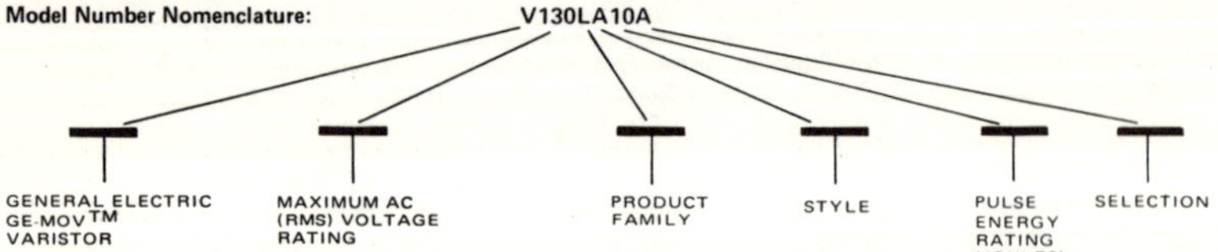

GENERAL ELECTRIC

Figure 3-4D. Metal oxide varistor specification sheet.

Maximum Electrical Ratings:

Maximum Energy, Power and Peak Current . . . . See Rating Table
Storage Temperature, T_{STG} . . . . –40°C to +125°C
Operating Surface Temperature, T_S . . . . 115°C
Operating Ambient Temperature (without derating) . . . . 85°C
Maximum Voltage Temperature Coefficient . . . . 0.05%/°C

Mechanical Ratings:

Insulation Resistance – Megohms . . . . >1000
Hipot Encapsulation – Volts D.C. for 1 Minute 2500
Solderability . . . . Per Mil Std 202D Method 208B

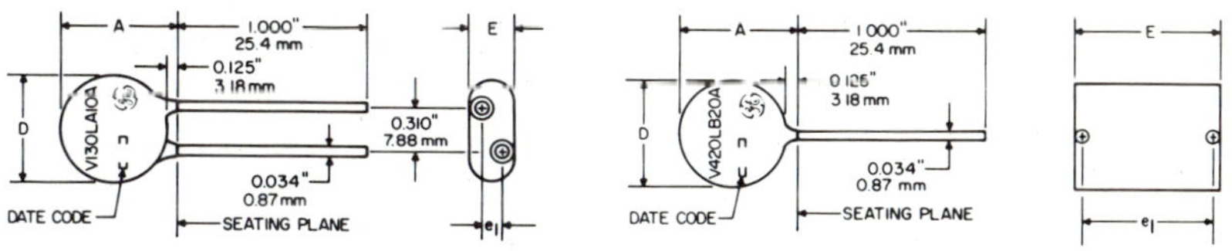

MODEL	A MAX. IN.	A MAX. MM	D MAX. IN.	D MAX. MM	E MAX. IN.	E MAX. MM	e_l MAX. IN.	e_l MAX. MM
V95LA7	.745	18.93	.620	15.75	.230	5.85	.110	2.80
V130LA10	.745	18.93	.620	15.75	.230	5.85	.110	2.80
V130LA20	1.005	25.53	.880	22.36	.230	5.85	.170	4.32
V150LA10	.745	18.93	.620	15.75	.230	5.85	.110	2.80
V150LA20	1.005	25.53	.880	22.36	.230	5.85	.170	4.32
V250LA15	.745	18.93	.620	15.75	.290	7.37	.170	4.32
V250LA20	.745	18.93	.620	15.75	.290	7.37	.170	4.32
V250LA40	1.005	25.53	.880	22.36	.290	7.37	.180	4.58
V275LA15	.745	18.93	.620	15.75	.290	7.37	.170	4.32
V275LA20	.745	18.93	.620	15.75	.290	7.37	.170	4.32
V275LA40	1.005	25.53	.880	22.36	.290	7.37	.180	4.58

NOTE:
1) Lead spacing dimensions as measured within 0.050 inches (1.27 mm) of seating plane
2) All dimensions shown are maximums except lead length which is a minimum

MODEL	A MAX. IN.	A MAX. MM	D MAX. IN.	D MAX. MM	E MAX. IN.	E MAX. MM	e_l MAX. IN.	e_l MAX. MM
V420LB20	.745	18.93	.620	15.75	.395	10.04	.345	8.76
V420LB40	.745	18.93	.880	22.36	.395	10.04	.370	9.40
V460LB20	.745	18.93	.620	15.75	.395	10.04	.345	8.76
V460LB40	1.005	25.53	.880	22.36	.395	10.04	.370	9.40
V480LB20	.745	18.93	.620	15.75	.395	10.04	.345	8.76
V480LB40	.745	18.93	.620	15.75	.395	10.04	.345	8.76
V480LB80	1.005	25.53	.880	22.36	.395	10.04	.370	9.40
V510LB20	.745	18.93	.620	15.75	.395	10.04	.345	8.76
V510LB40	.745	18.93	.620	15.75	.395	10.04	.345	8.76
V510LB80	1.005	25.53	.880	22.36	.395	10.04	.370	9.40
V550LB20	.745	18.93	.620	15.75	.700	17.78	.615	15.63
V550LB40	.745	18.93	.620	15.75	.700	17.78	.615	15.63
V550LB80	1.005	25.53	.880	22.36	.750	19.05	.640	16.26
V575LB20	.745	18.93	.620	15.75	.700	17.78	.615	15.63
V575LB40	.745	18.93	.620	15.75	.700	17.78	.615	15.63
V575LB80	1.005	25.53	.880	22.36	.750	19.05	.640	16.26
V1000LB80	.745	18.93	.620	15.75	.700	17.78	.615	15.63
V1000LB160	1.005	25.53	.880	22.36	.750	19.05	.640	16.26

FIGURE 1. DIMENSION TABLE

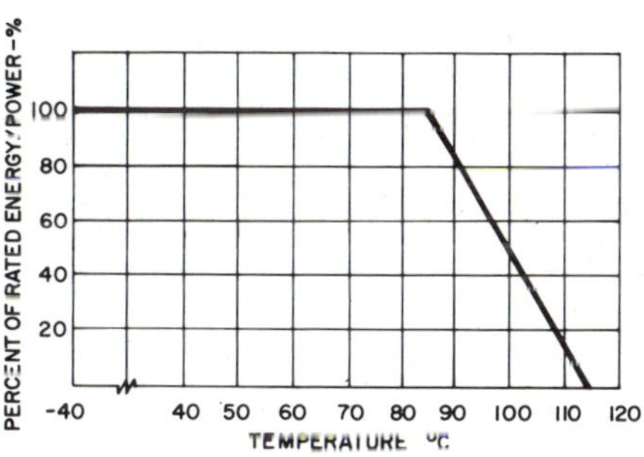

FIGURE 2. POWER AND ENERGY RATING VS TEMPERATURE

The average power input over the periodic time base resulting from successive voltage transients must be equal to, or less than, the GE-MOV varistor's rated average power dissipation at the specified ambient temperature! When this condition is met, the selected GE-MOV varistor has an energy rating high enough to suppress a voltage transient of the specified energy level at its operating surface temperature.

1. For higher average power dissipation, refer to General Electric's Power-MOV™ Varistor Series. Publication #180.67 3/73.

Semiconductor data handbooks are available showing the specifications and applications for an entire line of semiconductors. These books are jam-packed with information and some have as many pages as a telephone directory. If you are looking for a broad range of data on an entire company product line, you can't miss with one of these handbooks.

One hangup when writing to manufacturers is obtaining the correct address. The Howard W. Sams Photofact® Annual Index solves this dilemma. It includes a directory of addresses for manufacturers and importers of electronic home entertainment equipment. This handy alphabetical index makes short work of locating addresses.

3-5. HOW TO KEEP AN EFFECTIVE NOTEBOOK

Why try to store all that important information in your head? The chances are the knowledge that is not used regularly will slip away. Most electronic technicians are responsible for having access to a fantastic amount of information so make it easy for yourself by keeping a notebook.

The notebook system that is described in the following pages is easy to set up and use. Once you get in the swing of recording your data systematically, you'll wonder how you got along before.

Two notebooks are needed; a small shirt pocket size version and a larger 11″ × 12″ three-ring binder. Always keep the small notebook with you so it's convenient when you need it. If you have to hunt or walk across the shop for your notebook, the chances are it won't be used. When you're on the go, jot all useful data in the small notebook. Later at your convenience transfer this information permanently to the three-ring binder.

Figure 3-5 shows the large notebook. The three-ring binder is essential since it allows pages to be added or removed at a snap of a lever. In addition many data sheets and other worthwhile papers are prepunched for a three-ring notebook. The inside covers should have pockets for odd size papers. If you can't find one with pockets a little white glue and a manila folder or two will make them in a jiffy. Note the A to Z index guides shown in Figure 3-5. Dividing the notebook alphabetically simplifies organization immeasurably.

Figure 3-5. Three-ring binder notebook.

Often an electronic tidbit can be snipped out of an article and white glued in the appropriate spot in the notebook. It's a lot easier and faster than redrawing. In fact you'll probably jam the cover pockets with scraps and bits of information. Now and then take the time to white glue them in the correct alphabetical spot.

While we are discussing notebooks you might want to fasten a pack of colored fine line felt tip markers to the cover. These pens can't be beat when it comes to drawing schematics, showing electron flow, and circuit action. Besides, they will add a little color to your life!

4 Practical Shortcuts When Working With Hand Tools

Excellent hand tools used properly will reduce fatigue and make your work a pleasure. Good tools like fine wine seem to improve with age. This section will discuss worthwhile tools and techniques useful to the electronic technician.

4-1. SPECIAL PLIERS THAT WILL PAY FOR THEMSELVES

The most versatile tool for the technician is some type of pliers. Pick up a pair of pliers, place them in your hand and inspect them. Well-made pliers will fit your hand, feel good and have a nice balance. Open and close them; slight pressure from just two fingers should be all the force necessary. In fact, once the pliers are opened, the jaws should close easily by themselves just from the weight of the handle. Note the pivot joint and jaws while putting side pressure on the handles. No free play should be seen. Observe the jaws carefully. They should close evenly and mate each other uniformly. You won't be able to see light through the mating surfaces. The finish should be rust resistant. If the pliers you are holding fit the previous description they will likely give good service for a lifetime.

Figure 4-1A shows a workhorse 8″ long nose pliers. This tool is ideal for countless electronic applications. The handles are large enough for a comfortable and firm grip. You'll appreciate the extra length of these pliers when working deep into a chassis. There's plenty of leverage to make a hook connection using 16 gauge wire with little hand strain.

The 6″ pliers shown in Figure 4-1B is just what the doctor ordered for delicate IC or transistor work. The jaws are bent for easy access. They are extremely thin and fine, mating at the very tip. An excellent pliers for reaching into congested areas where the touch of a surgeon is required. Countless pliers of this nature have been ruined by rough and unthoughtful handling. This tool, like all fine tools, is made for one type of task. Use it for the proper application and it will never let you down.

The round jaws of the pliers in Figure 4-1C make wire bending and loop forming a breeze. Use these pliers when "square wiring" a chassis. The bends will conform to the diameter of the jaws. Need a small bend? Use the area of the jaw near the tip. For a larger radius bend use the jaw area nearer the handle. Bend resistor and other component leads with this tool to prevent the stress caused by sharp 90° bends.

A 6″ heavy duty combination long nose, side cutting pliers is pictured in Figure 4-1D. This multifaceted tool is just the ticket for everything from twisting wire around a terminal strip to holding a nut. The side cutting feature is especially convenient since a separate wire cutting tool is eliminated. This tool is built for everyday hard usage.

The side cutting or diagonal pliers "dykes" are found in every technician's tool box, Figure 4-1E. The diagonals pictured on the top are made so precise they will easily cut a hair. They are designed for cutting light gauge wire used for hook-up and component leads. Don't be like the technician who loaned his precision dykes out only to be told they wouldn't cut. The unthinking borrower had tried to cut piano wire leaving a permanent hole in the cutting edges. These dykes will cut a million transistor leads but one chop on a hardened wire ruins them forever.

At the other extreme, the rugged diagonals shown at the bottom of Figure 4-1E have been used for many years by the author for everything imaginable—and more; from pulling nails out of packing crates to cutting #8 gauge bolts. In spite of the hard and unusual usage they received, these dykes appear and work as good as when they left the factory. Confucius might say "He who buys quality tools spends the least money".

Did you ever have to cut close on a PC board but didn't

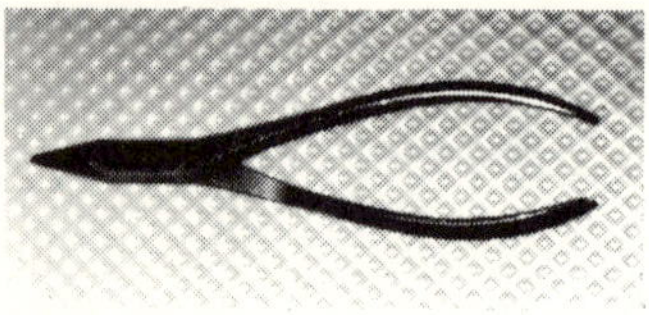

Figure 4-1A. 8″ long nose pliers.

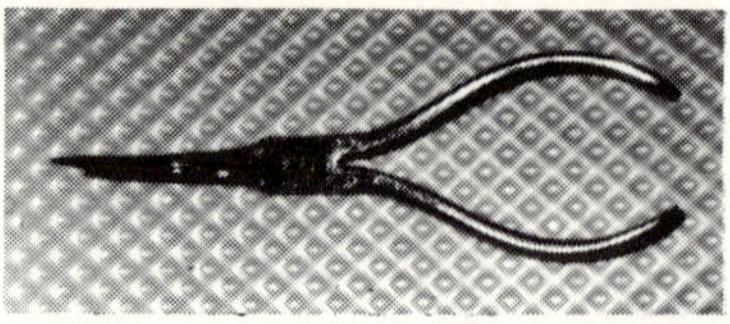

Figure 4-1B. 6″ bent nose pliers.

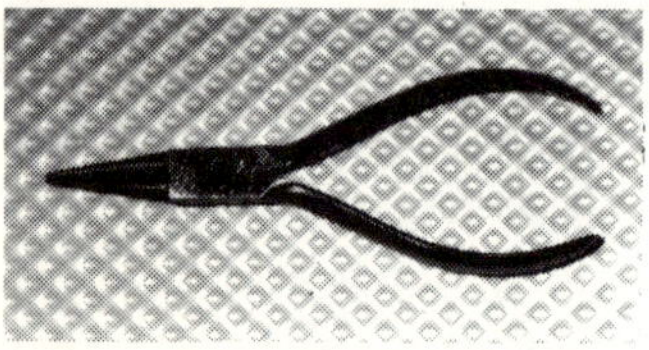

Figure 4-1C. 6″ round nose pliers.

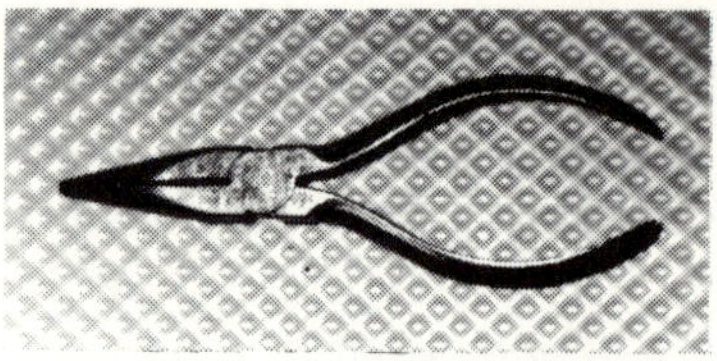

Figure 4-1D. 6″ heavy duty long nose pliers.

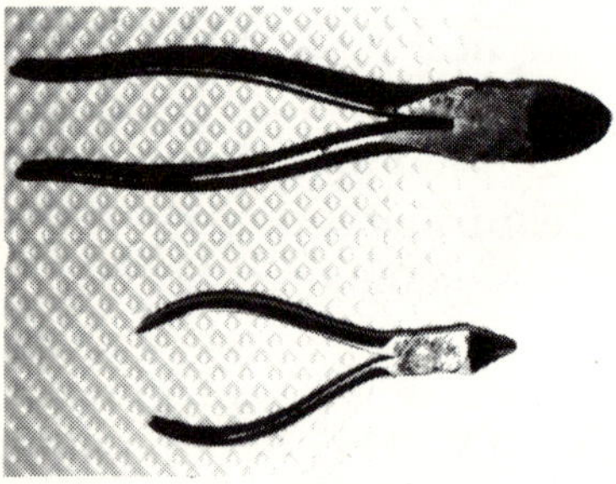

Figure 4-1E. Side cutting diagonal pliers.

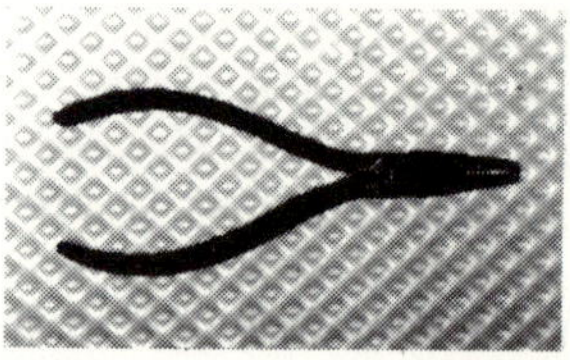

Figure 4-1F. 5″ end nippers.

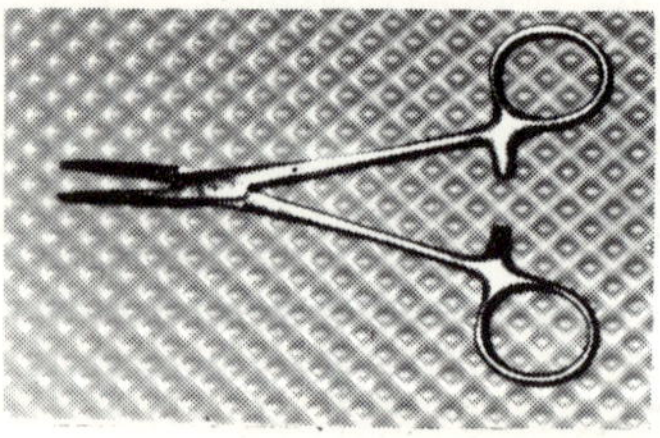

Figure 4-1G. Combination pliers-clamp.

have the space for dykes to fit? The end nippers, Figure 4-1F, are the answer. These little nippers have saved the day on many occasions. The cutting surfaces are built at the extreme ends of the jaws. If this baby saves removal of just one PC board for a part's change, it will have paid for itself. Figure 4-1G shows another dual purpose tool. This is a combination pliers and clamp with scissor type handles. Operating the handles with normal pressure gives long nose pliers action. Squeezing the handles causes them to lock the jaws tightly together. Everyone has needed a third hand from time to time; this is the solution. Use it as a heat sink, miniature vise, solder holder, nut starter, glue clamp, tool jig, or you name it.

The electronic technician like any craftsman can only be as good as his tools allow. The long nose and diagonal pliers described in this section are the heart of the tool kit. Select your pliers with care, use them wisely, and take pride in your work.

4-2. USEFUL TECHNIQUES FOR STRIPPING WIRE

Care should be taken when stripping insulation, so the wire is not cut or nicked. A cut on solid wire will weaken it, possibly causing a future break at the nick. Cutting too deep on stranded wire insulation will cut away many of the small strands reducing the current carrying ability as well as the loss of physical strength.

The wire stripper shown in Figure 4-2A is handy and inexpensive. It can be adjusted for various size wires by changing the screw and nut position on one of the handles. There are people who do not like this type of stripper because if not adjusted or used properly it will nick the wire. However, it is so compact and convenient that it is well worth the effort to learn how to use it correctly. Believe it or not there is a fair amount of

Courtesy of K. Miller T & M Co.

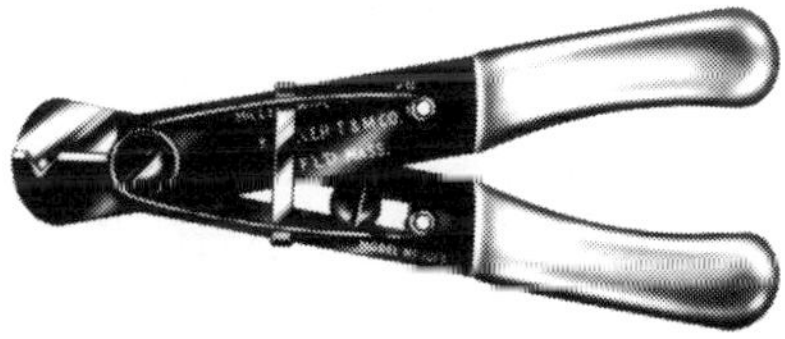

Figure 4-2A. Inexpensive wire stripper.

skill and manual dexterity necessary to become proficient with this stripper on all types of wire.

Practice the following technique and you'll never cut a wire. Set the adjustment for an opening slightly larger than the bare wire diameter. Cut into the wire insulation while rotating the stripper. If the adjustment is correct you won't be able to cut through more than three-quarters of the insulation. After the groove is cut around the insulation, remove the stripper placing it about 1/8″ from the original cut toward the wire end. Now use this new position to pull the insulation from the wire instead of cutting. The insulation will tear away from the original cut leaving the wire in mint condition. Once this technique is mastered, and it takes practice, you'll even be able to strip insulation from test lead wire and not touch a single strand.

One other little point to remember is to make it a habit to strip more insulation than necessary. It's a lot quicker to tuck in or cut off the excess wire than stripping over again.

There are numerous other kinds of wire strippers on the market including automatic and thermal types, but for prototype construction and servicing work the one shown in Figure 4-2A is tough to beat.

One of the easiest ways to open up a multiconductor cable is with a single edge razor blade. Figure 4-2B shows the procedure. Suppose you need the outside vinyl jacket removed 8″. Carefully score a ring around the cable 8″ from the end. Don't

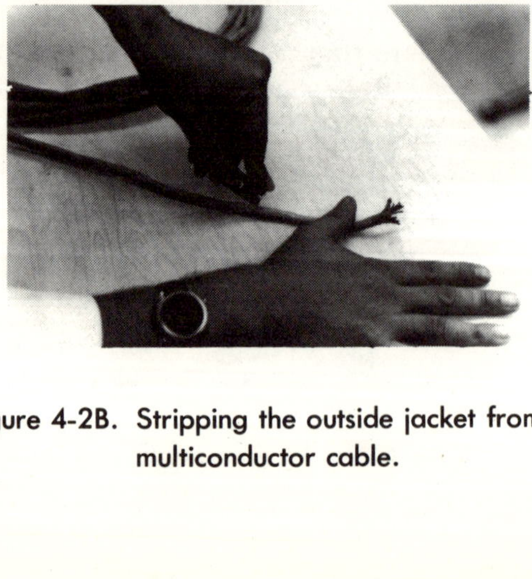

Figure 4-2B. Stripping the outside jacket from a multiconductor cable.

cut through the outside jacket since the inside conductors may be cut inadvertently. Now cut into the outside jacket at the cable end with the blade. Only cut through the jacket for about an inch. Reduce blade pressure continuing along the length of the cable, scoring a line until the blade reaches the scored ring. All that's left is to grasp the jacket at the cut end pulling back toward the circular score. The jacket will separate at the scored lines until detached.

Here's a method to professionally prepare the end of an insulated shielded coaxial cable. Remove the outside vinyl cover as detailed in the previous paragraph. The next step is to remove the inner conductor from the braided shield around it. Refer to Figure 4-2C for the following explanation. Push the shield down away from the end causing it to bulge. Now very carefully using a sharpened nail or scribe, work into an opening in the shield making the hole larger. Use caution so you don't break any of the shield wires. Double the lead over at the hole and very gingerly pull the center conductor out. Pull the braid back into its original length. When done properly it makes a beautiful job.

4-3. WHICH SOLDERING TOOL SHOULD BE USED

As a rule of thumb when only a few soldered connections are to be made use a quick heating gun; for a lot of soldering over an extended period use an iron. The soldering gun is most popular for service work where fast warm up is essential and only a part or two will be replaced. Electronic construction work calls for an iron. The soldering iron can be plugged in during the morning and be ready for service instantly throughout the day.

The electronic shop should have at least three types of soldering tool—gun, iron, and cordless iron.

A heavy duty dual heat, 240/325W, soldering gun is essential, Figure 4-3A. There are so many applications when fast, high wattage heat is needed. Any time a connection is soldered or unsoldered on a lug fastened directly to a metal chassis, a lot of heat is required to keep the chassis from absorbing the heat before the solder melts. In fact any tie point with multiple wire connections probably will solder best with the high wattage gun.

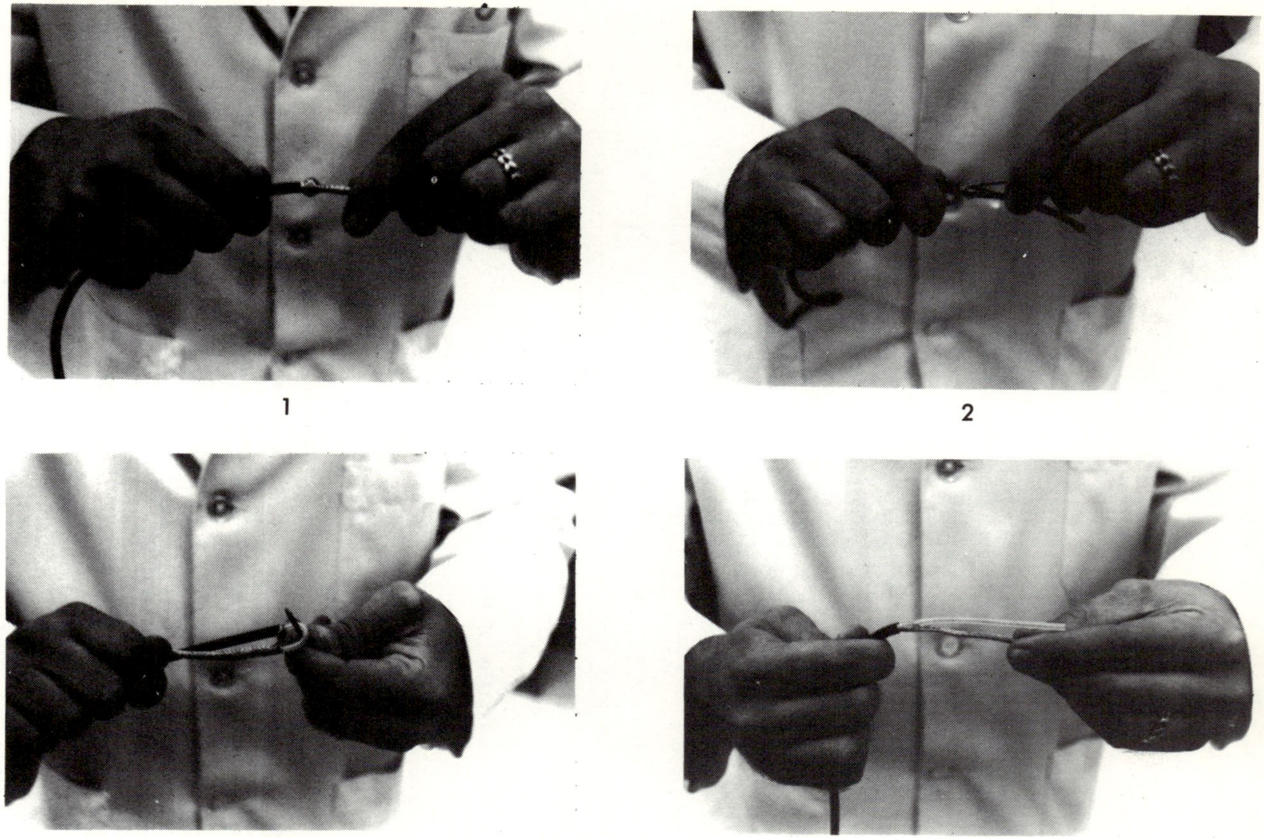

1 2 3 4

Figure 4-2C. Preparing shielded coaxial cable.

Courtesy of Weller Corp.

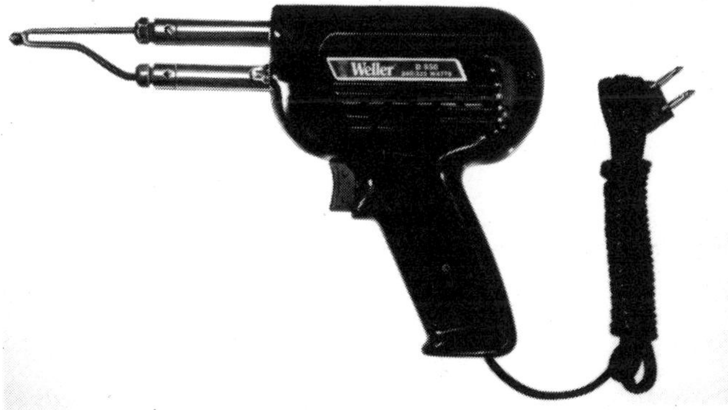

Figure 4-3A. Heavy duty dual heat soldering gun.

Use it to make solder connections directly to a chassis. If you ever try to remove a soldered UHF tuner cover without a high wattage gun you're in trouble. Most of these guns have small spot lights that are great for illuminating a dark chassis.

When choosing a soldering iron there are multiple factors to consider. First you must consider the heat of the tip which is determined by the iron wattage. The shape and size of the tip are also important. The iron should be rugged and be able to take a fair amount of abuse without breaking. It should feel good in your hand and have a nice balance. The handle must stay cool regardless of how long it has been operated.

There are irons made with temperature sensing tips that automatically maintain the tip at the proper soldering temperature. A 60 watt iron can do the work of fixed heat 15 to 100 watt irons. Another nice feature is that idling irons use minimum current. Most irons can accomodate various shape and size tips.

Three-wire soldering irons keep the tip at ground potential. This kind of protection is necessary to eliminate static induced currents when working with MOSFETs and other sensitive IC's. One note of caution; when using a three-wire iron, do not solder on a live or connected circuit. The tip is grounded and all kinds of fireworks will happen if you touch a hot line.

One constant problem with removable tips is seizure. Unless they are habitually removed, the tips will eventually seize to the iron making them almost impossible to remove without breaking. When the tip breaks you have got problems. Careful drilling through the center of the tip stub and removal with a screw extractor is usually the only solution. This is a very time consuming operation and is only about 75% effective. Routine use of anti-seize compound on the threads will help with this problem. Tips that screw on rather than screw in are much less likely to seize.

The newest types of soldering iron are the cordless, rechargeable models, Figure 4-3B, neat irons for portable and limited use. The tips usually heat in five or six seconds to about 700° F. A reasonable amount of soldering can be done before the battery is discharged; the number of solder joints you can make depends on the size of the connections and your soldering technique. It's smart to use the iron as soon as the tip melts solder. Much of the battery energy is used to heat the tip, so solder as many connections as possible each time the iron is used. Regular use of the iron is recommended for maximum battery efficiency.

Courtesy of Ungar, Division of Eldon Industries, Inc.

Figure 4-3B. Cordless soldering iron.

You'll like soldering with no cord to tangle or burn. When you're on a service call you won't have to hunt for an AC outlet or carry that long extension cord. You'll find the cordless iron so convenient that you will be using it for more and more routine applications. Small tips and built in illumination make them super for tiny and hard to reach spots.

4-4. DESOLDERING CAN BE EASY

There are numerous special tools built for desoldering. The desoldering tool shown in Figure 4-4A melts and removes solder at the same time. Simply heat the connection until the solder melts. Squeeze the vacuum bulb lifting the solder from the connection. This tool is so handy because it takes only one hand to operate leaving your other hand free to pull out the part.

Courtesy of Ungar, Division of Eldon Industries, Inc.

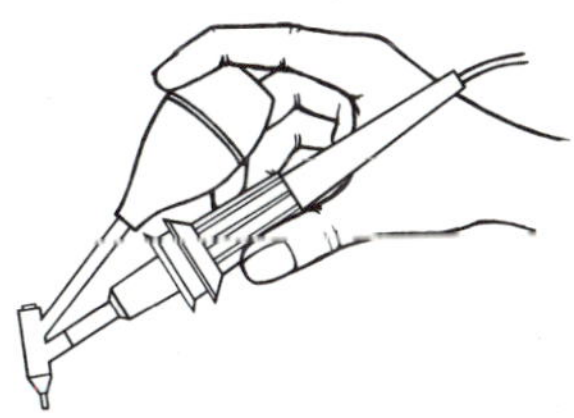

Figure 4-4A. Desoldering tool.

The vacuum bulb, Figure 4-4B, does a similar job of solder removal when used with your iron or gun. The only drawback is that both hands are used.

Courtesy of Ungar, Division of Eldon Industries, Inc.

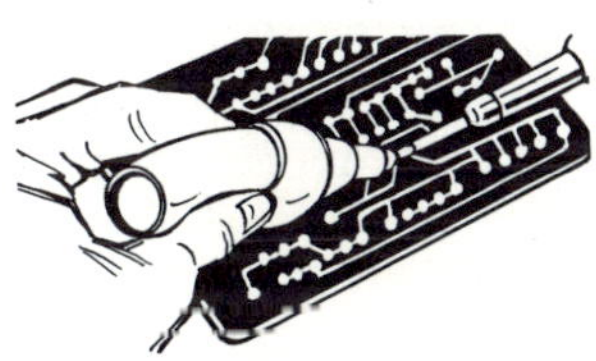

Figure 4-4B. Desoldering bulb.

Desoldering tips with specially shaped surfaces are a must for multiple-lead component removal, Figure 4-4C. Pick the correct shape tip for the component lead configuration. Before attempting to desolder make certain the desoldering tip is hot. It probably will take longer to reach solder melting temperature since most of these tips are a good deal larger than a conventional tip. Don't scrimp on solder when you tin the desoldering tip. In fact try to keep a puddle of molten solder completely covering the desoldering surfaces. This extra solder will compensate for irregularities in lead lengths and PC board surfaces, transferring the heat quickly from the tip to the component leads. If the tip is making good contact with all the pins, the component to be removed should be free in a jiffy.

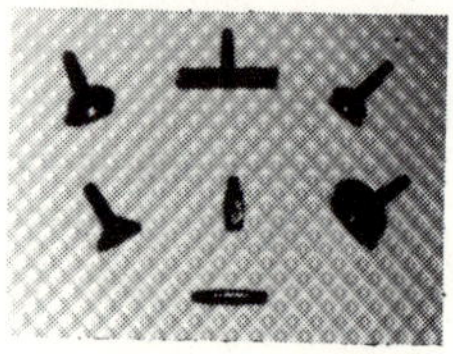

Figure 4-4C. Desoldering tips.

Probably the toughest components to extract from a PC board are dual in-line packages (DIP's). These IC's are something else to remove without breaking a lead or overheating. Fourteen or sixteen pins are a lot of leads to heat simultaneously. Fortunately there are tools made for removing these babies. Figure 4-4D shows a slotted bar desoldering tip and a spring loaded DIP extractor.

Courtesy of Ungar, Division of Eldon Industries, Inc.

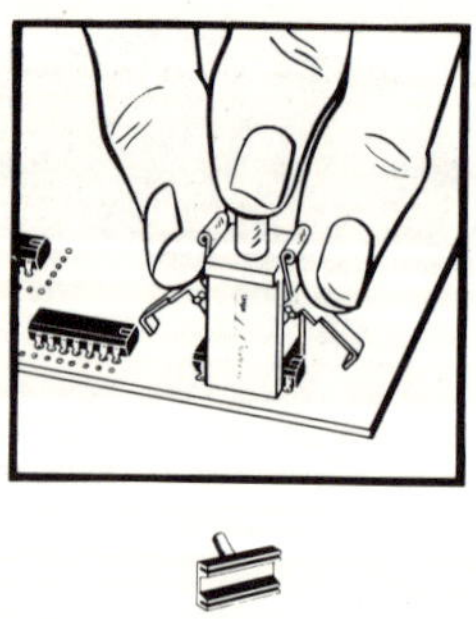

Figure 4-4D. DIP IC extractor tools.

Here's how they work. Clip the extractor on to the IC you are going to remove. Tin the slotted bar desoldering tip as previously described. As soon as the solder has melted on all the leads, the IC and extractor will jump off the board. It will really jump!

If you have many IC's or other components which are hard to remove, a small solder pot is the best answer, Figure 4-4E. This electrically heated pot holds several pounds of solder. Fill the pot to the brim; surface tension will keep the molten solder in the pot. Just set the foil side of the PC board on the pot over the IC. In a few seconds it will be free. You can remove a lot of sixteen pin IC's in short order with this method.

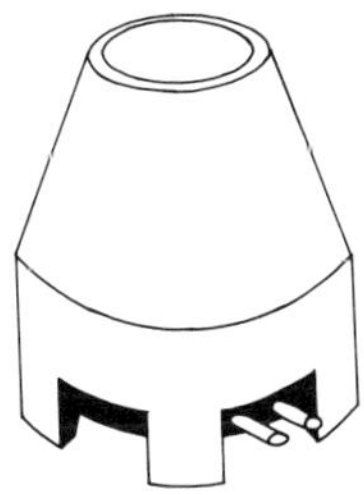

Figure 4-4E. Solder pot.

4-5. ALL ABOUT DRILLS AND FILES

Although drills and files are not normally associated with electronics, the electronic workshop certainly couldn't be without them. Number drills, 1 through 60, make up the most useful drill set. The drill index shown in Figure 4-5A conveniently holds this set of number drills. The largest drill, #1, measures .228″ and the smallest .040″. In addition to holding drills, drill indexes generally are stamped with some useful data:

1. Drill diameter
2. Common tap sizes
3. Tap drill numbers
4. Body drill numbers

Suppose you have to tap a 6-32 thread for a construction project. The drill index shows that #36 is the correct tap drill size. The body drill, hole size which a 6-32 screw fits through, is #28.

Figure 4-5A. Drill index containing number drills.

Other useful drill bits are 1/4″, 5/16″, 3/8″, 7/16″, and 1/2″. If the drill shank is cut down to 1/4″ the larger drills fit nicely into a standard 1/4″ portable drill chuck, Figure 4-5B. These larger drill bits are most useful when drilling front panel holes for chassis construction. Since you'll be drilling through some form of sheet metal most of the time, don't attempt to drill the finished hole with just one drilling. When sheet metal is drilled with a large size bit, the hole will be out of round. The answer to this problem is to drill the hole with a series of drills making each succeeding hole slightly larger than the previous one. If you don't want to bother drilling a series of holes, grind the larger drill points as shown in Figure 4-5C.

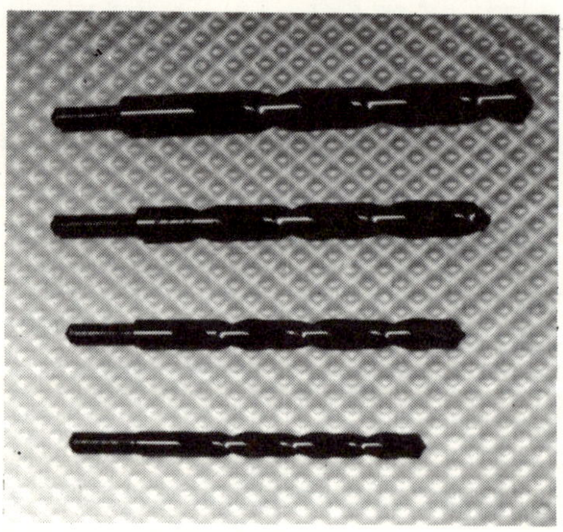

Figure 4-5B. Large drill bits with cut down shank.

Figure 4-5C. Drill point ground for cutting sheet metal.

After sheet metal has been drilled there are always burrs left around the holes. The quickest way to deburr small holes is to hand twist a counter sink or larger drill bit in the hole. The burrs will shear off as pretty as you please. You can make a good deburring tool for larger holes using a worn out three corner file, Figure 4-5D. Hollow grind the three flat surfaces on a grinding wheel. Grasp the deburring tool by the handle and cut around the circumference of the hole like peeling an apple. One time around will usually cut all the burrs.

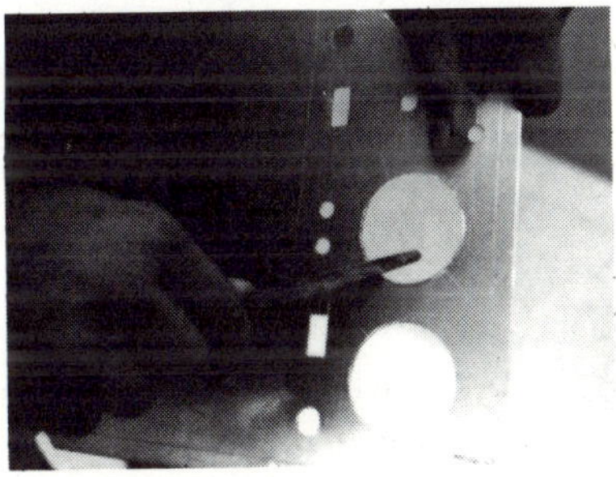

Figure 4-5D. Deburring tool for large sheet metal holes.

An assortment of files most useful for electronic construction and repair are shown in Figure 4-5E. A double cut, second cut tooth structure generally works well for sheet metal. The hand file shown can be used for filing straight edges, cleaning up and straightening out square and rectangular holes. Use the curved side of the half-round file for enlarging or touching up meter holes. The round file is invaluable for elongating or enlarging smaller holes. Small rectangular openings like those needed for slide switches can be made with the square file.

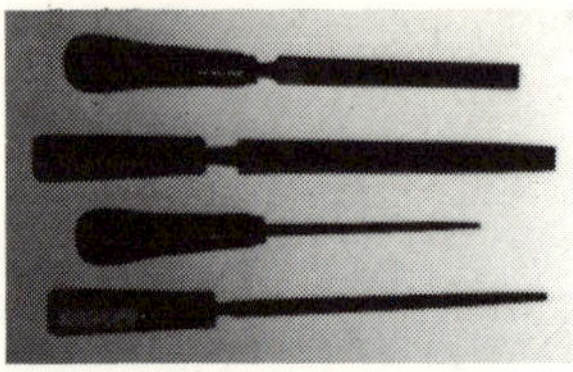

Figure 4-5E. File assortment for electronics.

When you want to obtain a smooth finish on an exposed chassis edge use the technique called draw-filing, Figure 4-5F. The hand file is moved back and forth perpendicular along the chassis edge with uniform pressure. Teeth marks or other edge irregularities will be speedily removed.

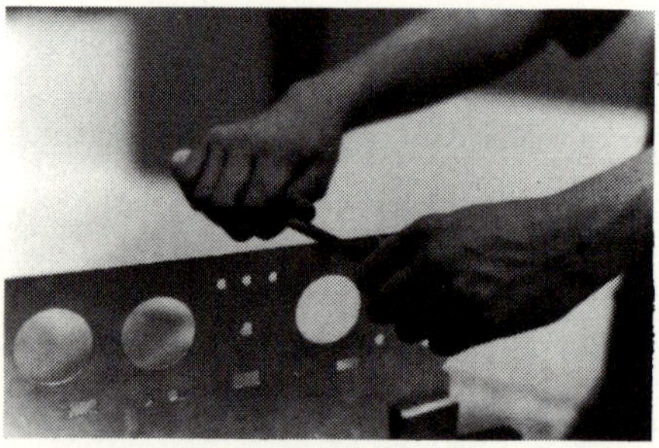

Figure 4-5F. Draw-filing a chassis edge.

You'll need a set of jeweler's files or swiss pattern files for delicate work. The set shown in Figure 4-5G will take care of just about all electronic applications.

The following 4" files are included in the set:

1. Round
2. Square
3. Triangular
4. Half round
5. Flat
6. Hand

Remember that fast and accurate metal filing requires a great deal of skill. It's a wise policy to work more slowly while continually checking your progress. A filing mistake is usually difficult to cover up once the metal has been cut away.

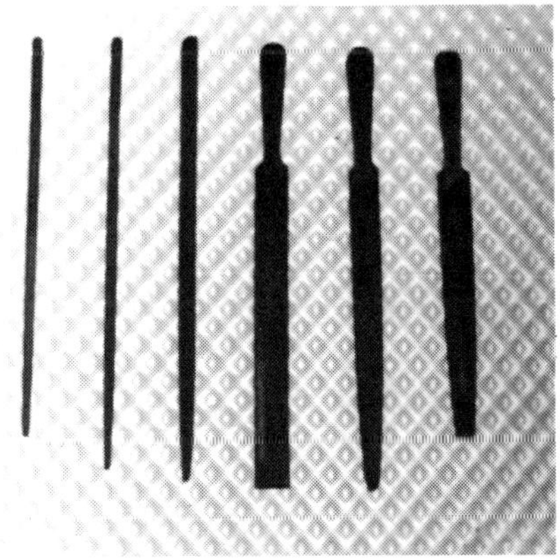

Figure 4-5G. Jeweler's file set.

5 Effective Ways To Use Essential Test Instruments

When it's all said and done, the fact remains that in electronic servicing work surprisingly few key test instruments do most of the work.

This chapter will deal with these workhorse types of test equipment. Some interesting tips for working with VTVM's will be discussed and you will read about methods for tracking down electronic bugaboos with audio and RF generators. Answers to your questions concerning tube checkers will probably be found in the following pages. In the last section, you'll find timesaving techniques involving the use of a multiple component substitution box.

5-1. WORKING TIPS WITH THE VTVM

Most technicians use a VTVM regularly and are familiar with the common functions; AC, −DC, +DC, and ohms. However it's common to overlook some VTVM features that are not routinely needed.

For instance, many VTVM's have scales that can be read directly in peak to peak AC voltages. This direct reading feature saves time as the usual AC RMS readings do not have to be multiplied by 2.8. However the P to P scales are only accurate for sine waves. The meter will give an indication for pulse and other unsymmetrical waveforms but the reading will correspond to the average voltage and not true peak to peak, see Figure 5-1A.

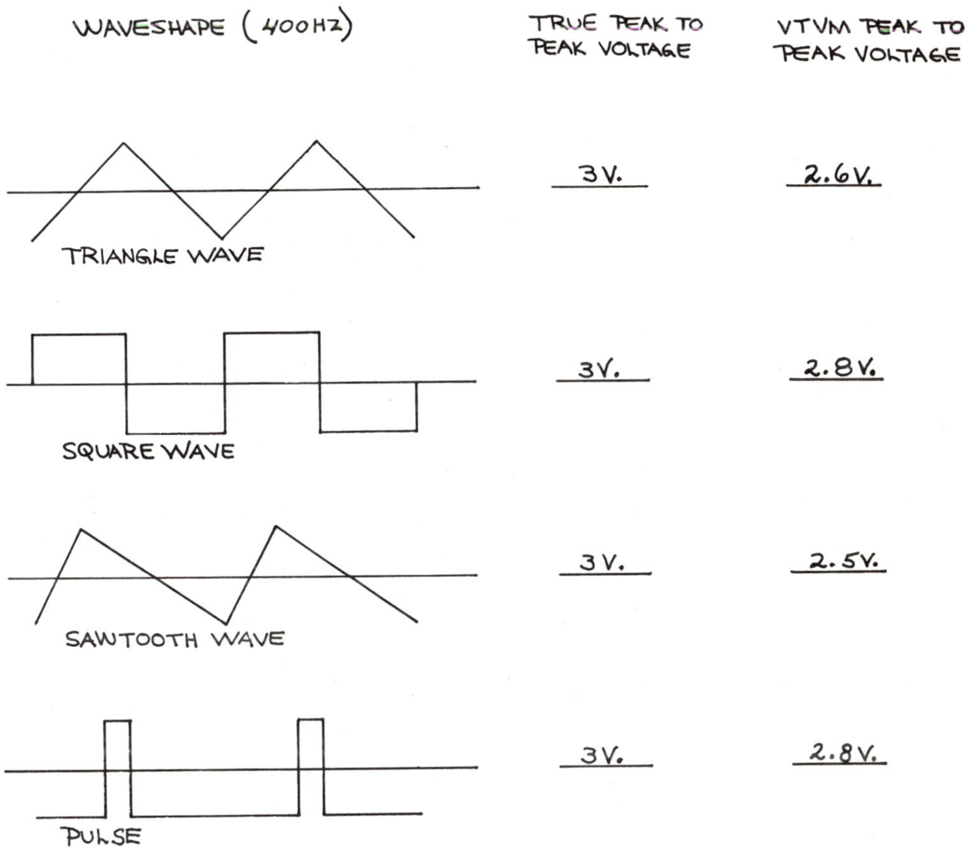

Figure 5-1A. Peak to peak VTVM voltages for various waveshapes.

Another convenient VTVM feature is the center zero option. The zero adjust control can be set so the pointer rests at center scale when the probes are shorted, Figure 5-1B. This option is especially helpful when performing alignments calling for a positive and negative voltage swing around a zero null.

Most VTVM meter faces have a zero ± dB scale. The human ear does not respond to sound volume in proportion to signal strength so a unit of measurement nearly equivalent to human ratios is needed. The common unit for signal strength is the decibel (dB).

The dB scale is usually marked for the lowest AC volt scale. Each time the AC range switch is increased a position, you must add 10 dB to the reading. For example if the pointer was at +3 dB and the AC range switch was clicked to the third range, the indication would be +23 dB.

Figure 5-1B. VTVM set for center zero.

Figure 5-1C shows how to set up a VTVM dB measurement for an amplifier frequency response check (also refer to Section 5-2). Adjust the audio generator for some reference frequency. 1,000 HZ is common. Set the gain of the generator until the VTVM reads zero dB. Now vary the frequency of the generator through the amplifier range while monitoring the VTVM; dB indication above and below the reference frequency will be clearly shown.

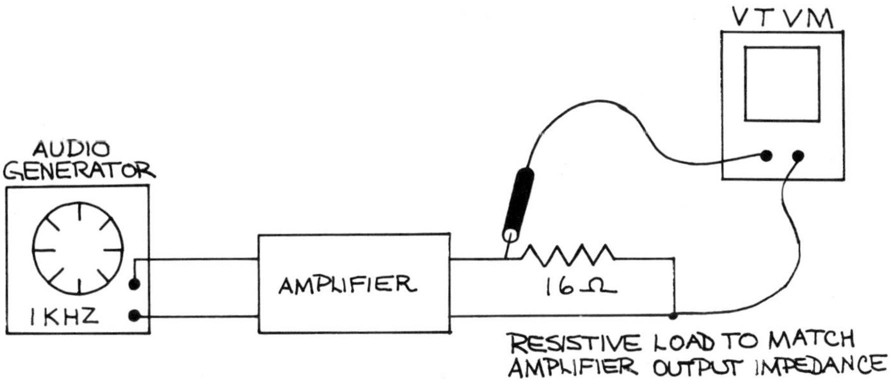

Figure 5-1C. VTVM dB measurement.

Four technicians individually were asked to check the forward and back resistance of a silicon rectifier using the ohmmeter section of one VTVM. Here are the results:

Technician	Forward Resistance	Back resistance
1	11 ohms	2 meg ohms
2	90 ohms	infinity
3	600 ohms	infinity
4	5,000 ohms	infinity

Technicians 1 and 2 said that the rectifier was okay. Technicians 3 and 4 rejected the rectifier, stating the back resistance was good but the forward resistance was too high for a silicon rectifier.

What's the story? The rectifier and the VTVM were the same for all the technicians. They all felt they could check a rectifier. Each knew how to zero and ohms adjust the meter as well as read the scales. Any time a VTVM is used to measure the resistance of any solid state junction, the junction resistance is determined by the ohmmeter scale being used. Technician #1 used the proper technique. The forward resistance, 11 ohms, was measured on the lowest ohms scale, R x 1. The ohmmeter polarity was then reversed and the ohms range switch increased until the back resistance could be measured. The second technician used R × 10 scale for forward resistance and didn't turn the range switch high enough to detect the 2 meg back resistance. The third and fourth technicians checked back resistance like the second but used R × 100 and R × 1,000 positions, respectively, for their forward checks.

The rule for ohmmetering solid state junctions: use the lowest ohmmeter scale for forward resistance measurements and the highest scale necessary to obtain a reading for the back resistance. Naturally the ohmmeter polarity is switched between forward and back measurements. The only time this rule should not be observed is when the higher currents of the R × 1 range may damage a delicate solid state junction. The vast majority of solid state devices are not harmed by the R × 1 range.

5.2. SUCCESSFUL HIGH-FIDELITY AMPLIFIER TESTS WITH THE AUDIO GENERATOR

Let's begin by making a frequency response check of an audio amplifier using sine-wave analysis. Set your equipment as shown in Figure 5-2A. For the most accurate measurements the speaker should be substituted with a non-inductive power resistor which is the same resistance as the speaker inpedance. Also make sure the resistor wattage is sufficiently high to handle the

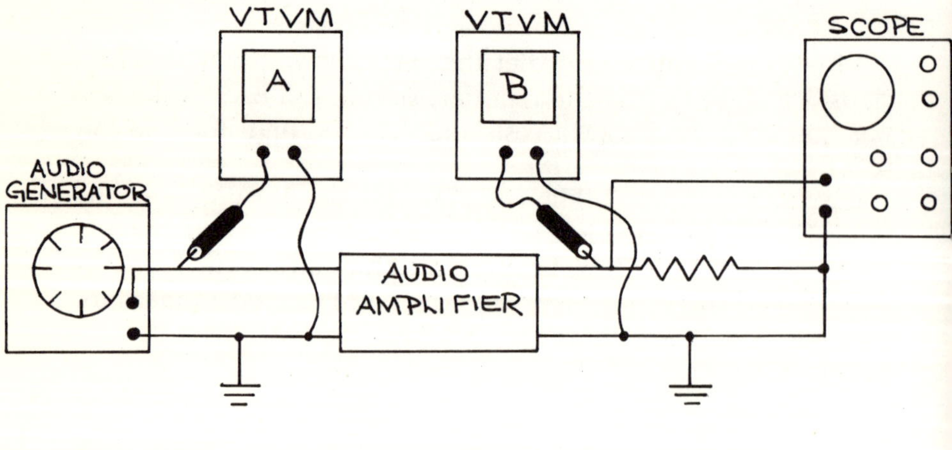

Figure 5-2A. Equipment setup for making a frequency response check.

amplifier power. Set the amplifier volume control for maximum and the tone controls to their neutral positions.

Adjust the generator for a reference frequency (1,000 HZ). Increase generator gain while monitoring the oscilloscope. When the 1K HZ sine wave begins to distort and peaks flatten, back off on the gain to the point just before distortion. Set VTVM A for center zero (0 dB), see Section 5-1. Note, it's important to maintain VTVM A at this reading for all test frequencies by changing the gain of the audio generator: a good generator should provide constant output regardless of frequency. Also set VTVM B for center zero (0 dB).

Tune the audio generator to the lowest frequency of the response check. Record the ± dB reading indicated on VTVM B. Repeat this procedure for the remaining frequencies using the following system.

1. 10 HZ steps until 100 HZ
2. 100 HZ steps until 1,000 HZ
3. 1,000 HZ steps to the amplifier high frequency limit

Plot the points on logarithmic graph paper similar to the response curve shown in Figure 5-2B. Repeat the previous procedure for a moderate audible level by backing off on the volume control to the desired position. After plotting the second curve repeat the procedure again for a low audible level. These three response checks at the various power levels will provide an accurate analysis of the amplifier.

Perhaps you would like to make a quick check of an audio amplifier but are not particularly interested in the full response curve. Here's a fast method to make a frequency response check using square waves. Many audio generators can be changed to a square wave generator with a flick of a switch. A square wave consists of a fundamental frequency and an infinite number of odd harmonics. So, when an amplifier passes a square wave with little or no distortion it simply means that the fundamental and a whole army of harmonic frequencies were amplified with no distortion. Refer to section 13-2 for a detailed procedure explaining this technique.

Figure 5-2C shows how to determine the approximate impedance of a high fidelity amplifier. To measure the input

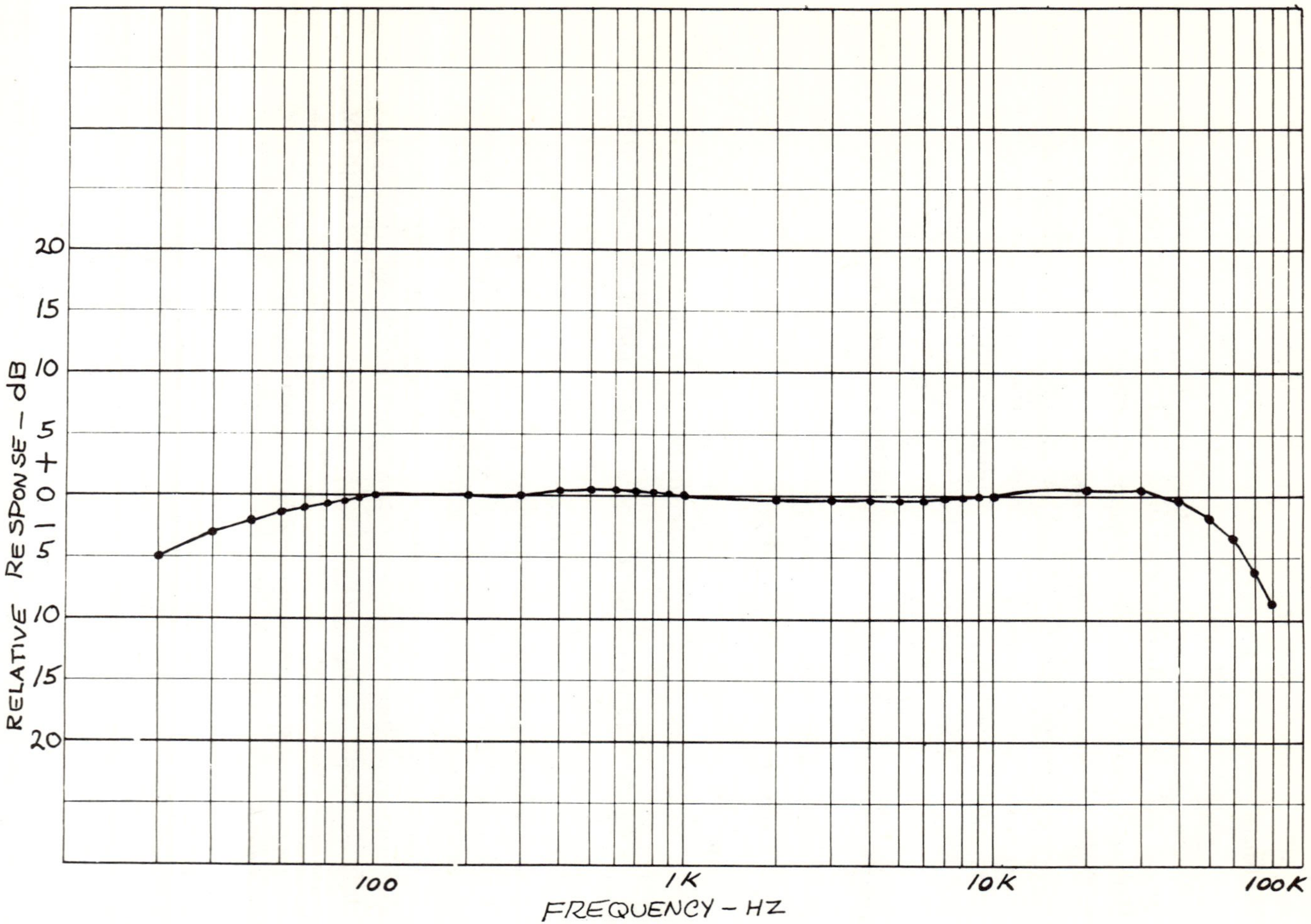

Figure 5-2B. High-fidelity amplifier frequency response curve.

impedance of an amplifier, inject a low 400 HZ signal into the amplifier input through a potentiometer. Turn the pot until the VTVM reads maximum AC voltage. Now change the pot until the VTVM reads onc half of the maximum voltage. Remove the pot from the circuit and measure the resistance between A and B of the pot; the resistance measured equals the amplifier input impedance.

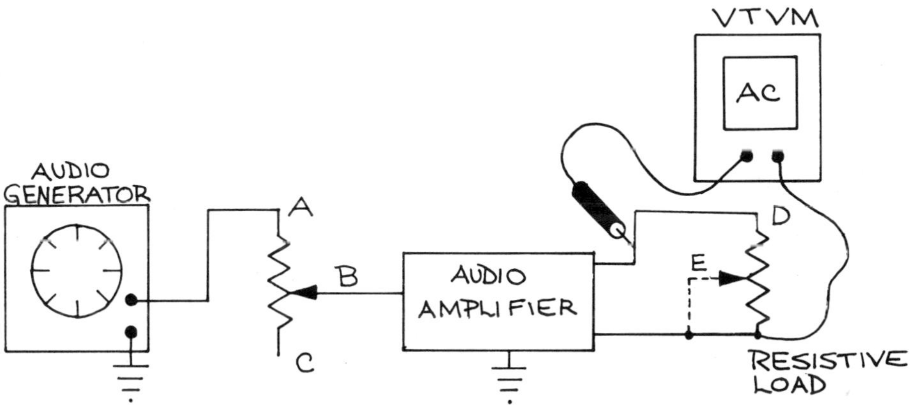

Figure 5-2C. Determining the input and output impedance of an audio amplifier.

To measure the output impedance connect a potentiometer (dotted lines) as shown in Figure 5-2C. Adjust the output pot until the VTVM reads one half of the original voltage. The resistance between D and E is approximately equal to the output impedance. Since some output impedances are inductive be sure to check the output impedance over a range of frequencies. Needless to say, the potentiometers for these tests should be non-inductive, approximate resistance should be close to the amplifier impedance, and the appropriate wattage.

5-3. USE THE RF GENERATOR FOR QUICK, EASY TV TESTS

An invaluable aid for servicing video and sound troubles in a television receiver is the RF generator. This basic test instrument is commonly used for all kinds of electronic alignment jobs but often overlooked when troubleshooting TV's. The next few paragraphs will describe simple sure-fire ways to locate video, IF, RF, and FM sound related problems.

For the following discussion let's assume that the TV described has been operating normally when the trouble occurs. The trouble symptoms will be video, sound, or video and sound related problems. The picture tube is lit with normal brightness (good raster), the associated tubes or plug-in transistors have been checked okay, and the power supplies are good.

The first case is a portable TV with good sound, good raster, but no video. Reach for the RF generator and you are ready to find the defective stage. Most RF generators have a separate 400 HZ audio output jack with the gain varied by a front panel control. Provide a common ground between the generator and the TV. Apply the full audio output signal to the video input of the CRT, see Figure 5-3A, point 1. If the CRT and associated circuits are working properly you should see six grey horizontal bars on the picture tube face as shown in Figure 5-3B.

Next move the 400 HZ signal to point 2. The bars should become darker indicating amplification in the video output stage. No bars or washed out bars indicates trouble in the video output. If the video output checks out move the generator signal to point 3. The normal CRT indication will be six blacker horizontal bars. A defective video amplifier would stop the bar pattern or decrease its contrast. Since the sound is normal, the problem will show up in some stage between the video amplifier and the picture tube.

The second case to analyze is a TV having no sound, no video but good raster. Refer to Figure 5-3C during this explanation. Since the sound and picture are both missing, the problem area will be a stage that is common to both signals, the IF's or RF's. Turn the TV channel selector to an *unused* channel in the high end (13) of the VHF band. Tune the RF generator to the IF picture carrier frequency, usually 45.75 MHZ. Note, some generators may not produce a signal capable of reaching this high frequency. In this case dial the generator to one half of the 45.75 MHZ and the second harmonic will do the job. Feed a 400 HZ modulated IF signal to point 1. If the IF's are working you'll see 6 black horizontal bars on the picture tube. Inject the signal into point 2 and then into point 3 if still no bar pattern appears. When the bars appear the stage to the left of the test point will be the trouble area.

If the IF's pass the bar signal the defect must be in the tuner. Next inject the modulated IF signal into the mixer input, point A. The bar pattern will again be seen clearly on the CRT if the mixer is working. The only two stages remaining are the RF oscillator and amplifier. Try feeding a modulated RF signal into the input of the RF amplifier, point B. This stage is controlled by the channel selector so make sure the generator fre-

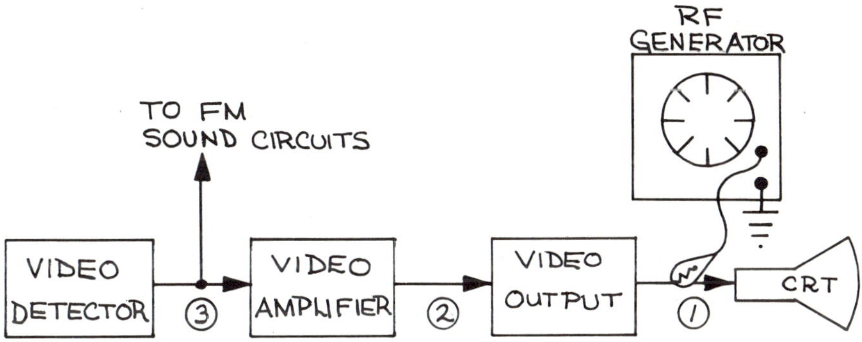

Figure 5-3A. Troubleshooting a defective video section using an RF generator.

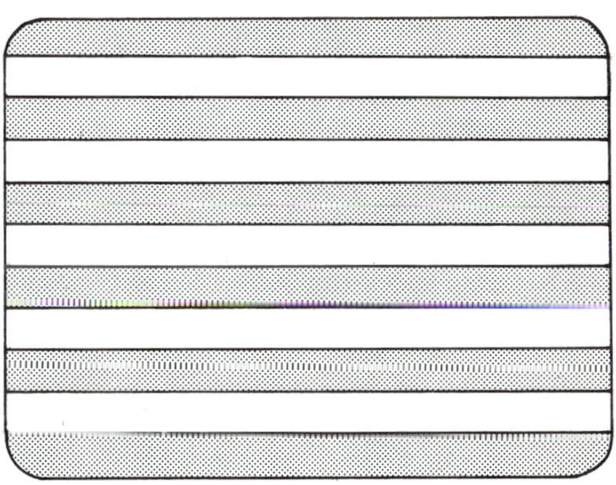

Figure 5-3B. 400 HZ CRT bar pattern.

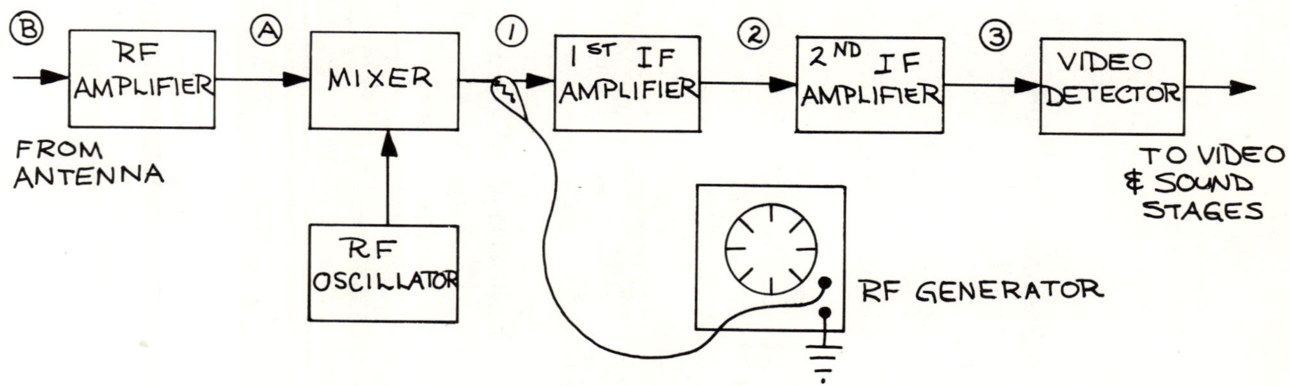

Figure 5-3C. Troubleshooting a defective IF or RF stage using an RF generator.

quency matches the channel frequency. For instance, the picture carrier frequency for channel 2 is 55.25 MHZ. Set the generator for this frequency or one half of that value if the generator doesn't go high enough (the second harmonic). If you see the CRT bar pattern, the RF amplifier and oscillator must be working. When no bars are present shift the generator lead to the RF amplifier output. No bars probably indicate a bad oscillator; bars mean the RF amplifier is defective.

Once the RF generator and bar pattern techniques are mastered it is a very quick procedure to check out TV video circuits. In summary, use 400 HZ signal for video circuits, 400 HZ modulated picture IF frequency for IF's and mixer stages, and 400 HZ modulated channel selector frequency for RF circuits. As each stage amplifies the bar pattern, the bars should become darker. Keep decreasing the gain of the generator to prevent overloading and false indications as more amplifying stages are being used. Before trying this troubleshooting procedure on a defective receiver, it would be most helpful to practice on a working TV. It will give you an idea of what to expect in generator gain and bar pattern contrast for the various injection points.

The third TV to repair has good video and raster, but no sound. The trouble must be in the sound chain. Identify whether the trouble is in the audio or FM circuits by feeding a 400 HZ audio signal into the top of the volume control. A good tone from the speaker indicates FM sound trouble. No tone means audio problems. Troubleshoot the audio circuit in the manner described for video circuits. Use speaker tone in place of the bar pattern for this procedure.

Here's how to check the FM circuits if the audio section is okay. Turn the channel selector to an unused channel. Connect a VTVM set for zero center across the FM detector load resistor, see Figure 5-3D. Adjust the generator frequency for the sound IF center frequency, generally 4.5 MHZ and inject into point 1. Turn the generator dial back and forth around 4.5 MHZ. If the second sound IF and FM detectors are working, the VTVM will swing above and below center zero with the change in generator frequency. Move the generator to point 2 to check the first sound IF in the same manner. Once the defec-

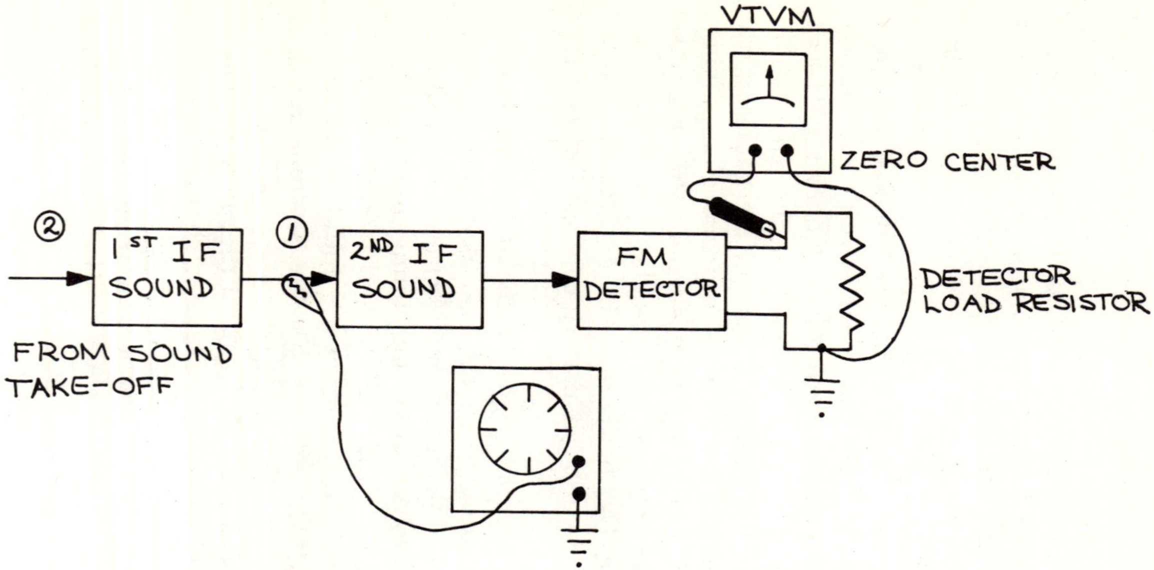

Figure 5-3D. Troubleshooting a defective FM sound section using an RF generator.

tive stage is located, check it carefully with voltage and resistance measurements to find the bad part or parts.

5-4. SHORTCUTS FOR WORKING WITH TUBE CHECKERS

Tube checkers are invaluable test equipment when operated by an experienced electronic technician. All tube checkers have limitations that must be understood before meaningful conclusions can be made.

Emission and transconductance testers are the two basic types on the market. The emission type shows cathode emission of each tube connected as a diode; each grid is connected in parallel with the plate. The more expensive transconductance checkers operate the tube as an amplifier with appropriate grid, plate, and screen voltages.

High frequency, high voltage, high current and oscillator tubes operate normally under conditions that tube checkers can't duplicate, Figure 5-4A. Any tubes that fall into one of these categories must be questioned and tested. Naturally if any tube tests dead, showing no emission or conductance, or if any element shows a short or open, the tube should be discarded.

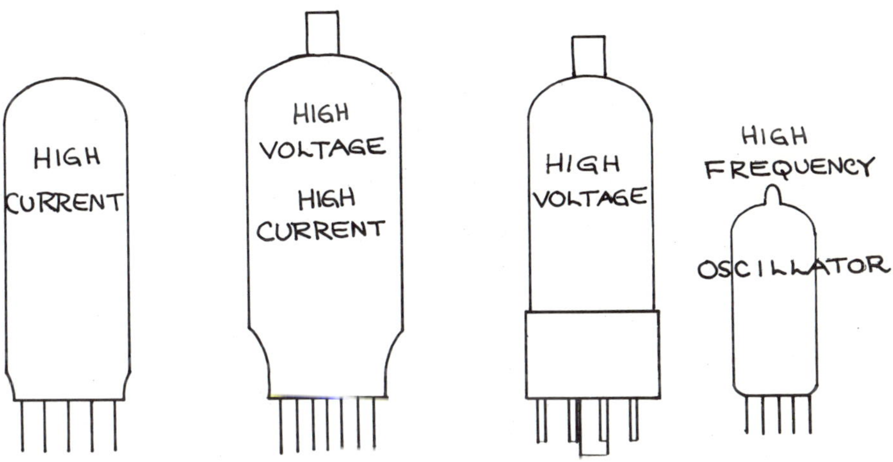

Figure 5-4A. Tube checker's nemesis.

The first test to be made in any type of tube checker should be for shorted elements. The reason is that a shorted element can peg or damage the meter movement if testing is continued. While short testing tap the tube with a common lead pencil around the envelope. Many shorts only show up with a little mechanical stress. Some shorts appear only after the filaments have fully heated the tube. Don't rush the short test because you are likely to miss shorts caused by thermal stress. Sometimes a short indication is evident but disappears with tapping. Don't assume the short's gone forever. Discard the tube or it will come back to haunt you sooner or later.

If the tube checker panel lamps dim noticeably and do not return to their normal brilliance very quickly after a tube is inserted, pull the tube immediately. The tube probably has a plate-to-filament short which could easily wipe out the power transformer. Tube checker's power transformers are very costly in addition to having a million leads to connect.

When no shorts are evident, next check for quality (emmission or conductance). Here's where psychic powers really pay off! Hope that the meter pointer doesn't budge. Then you are positive the tube's defective. If the meter shows that the tube is in the acceptable area it may be good or bad depending on the kind of tube and circuit. The tubes to be most suspicious of are shown in Figure 5-4A. However, it's remotely possible for any tube to check good and not work in the actual circuit.

Most tubes that show quality just under the acceptable region will operate as well as those that show full quality. Even tubes in the "replace" category will often work without a noticeable deficiency. However if you're checking a string of amplifier tubes such as those found in an IF section and they all check weak, the cumulative effect will probably be noticeable. While checking emission, tube tapping should not change the meter reading; discard the tube if the reading changes.

Almost all modern tubes are multiple tube types containing more than one set of elements. It's easy to forget to check a section when you're in a hurry. Make it a habit to check the roll chart for all listings before you pull out the tube. There's nothing worse than thinking you have checked a tube when in fact a section has been overlooked.

Of course there are some tube defects that are visually apparent, Figure 5-4B. The most common is when the filament is simply burned out, producing no light or heat. A tube with a whitish cast inside the envelope means air has entered, usually from a glass fracture. It's definitely bad. Arcing between tube elements is a sign for a new tube. Occasionally a tube will show signs of heating, warping, or melting on its plate surfaces. The tube is shot, but before replacing it, determine why it happened.

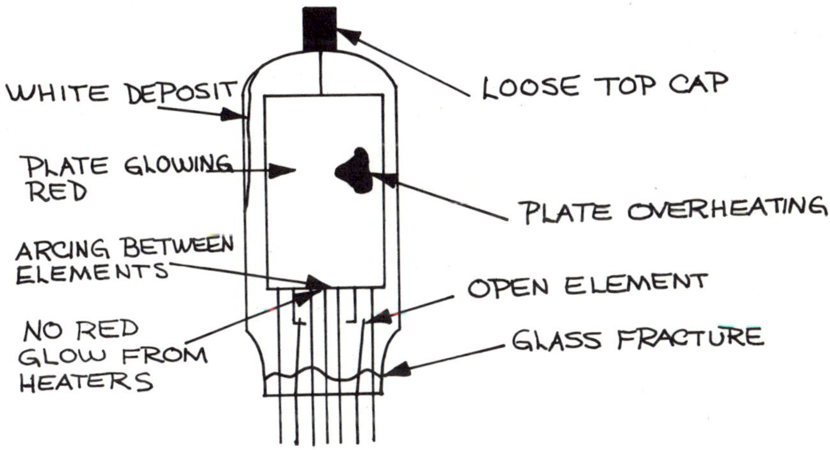

Figure 5-4B. Common visual tube defects.

The only 100% reliable test for a tube is substitution with another known good tube—a tube that you know has actually operated in that particular circuit. Of course it's impossible to have substitute tubes for all contingencies, so . . . the tube checker.

5-5. TIMESAVING TECHNIQUES INVOLVING THE MULTIPLE SUBSTITUTION BOX

No electronic workbench should be without a multiple substitution box. A large selection of resistor, capacitor, and rectifier substitutes are available just by turning knobs. Figure 5-5A shows a deluxe R-C substitution box that is especially handy. This unit is so versatile because each section works independently. For instance four different parts could be substituted simultaneously; a carbon resistor, wire wound resistor, capacitor or rectifier, and an electrolytic capacitor. That's convenience!

Courtesy of Sencore

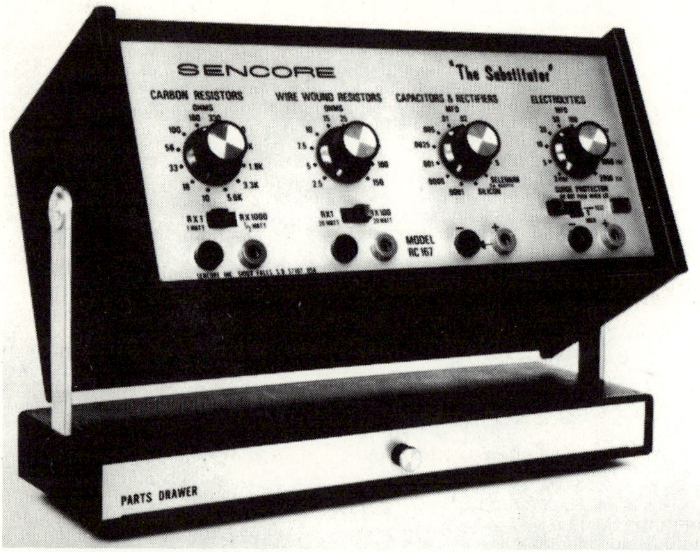

Figure 5-5A. R-C-rectifier substitution box.

Suppose you're repairing an FM stereo that has a hum problem on both channels. The hum is present even when receiving no station signal and the volume controls are turned down. These three clues point to a power supply filter problem. One technique would be to fire up your oscilloscope and measure the ripple voltage across the filter capacitors, comparing your measurements with the service data. The capacitor which has abnormally high P to P voltage across it will be the likely culprit.

The previous example is a good troubleshooting procedure. However, the substitution box can also find that bad filter capacitor and probably in a lot less time. Since most power supplies must filter low 60 HZ frequency, large electrolytic capacitors are necessary. Electrolytic capacitors have a tendency to dry out and lose much of their capacity. What could be simpler than selecting the approximate value electrolytic capacitor on the substitution box and clipping it across the filter capacitors one at a time. Do this while the stereo is operating and be sure to observe capacitor polarity. As soon as the defective capacitor is bridged the hum will disappear like magic. You've pinpointed the trouble in less time than it would take the scope to warm up.

Now you are working on a cherished, older model shortwave radio. The radio appears to be dead although all the tubes have checked out and their filaments are lit. A quick check with a VTVM reveals low DC plate voltage on all tubes. The same low B+ is present at the output of the rectifier. Look at that rectifier again, it's selenium. Selenium rectifiers are noted for increasing their forward resistance with age. Select an equivalent selenium rectifier from the substitution box and clip it in parallel with the radio's selenium; anode to anode, cathode to cathode. If Radio Moscow comes blaring out of the speaker you know the trouble has been diagnosed.

So much for troubleshooting. Let's design an IC dual tone alarm. Figure 5-5B shows the circuit using the 556 double timer. Each half of the 556 timer is wired as a multivibrator circuit, and they work together to produce a two tone sound. Now this circuit can produce about as many dual tones as you would ever want merely by changing the timing circuits; just the ticket for the multiple substitution box.

Connect the capacitor section in place of the .2 MFD capacitor, and the wire wound resistor section instead of the 1K. Replace the 1 meg resistor with the other resistor section. Apply power while listening to the weird sounds that can be made by turning the substitution box knobs. You might want to substitute the .1 MFD capacitor for even more variations. The multiple substitution box makes it a snap to try all sorts of RC combinations for the ultimate sound.

Refer to the power supply circuit shown in Figure 5-5C. Looks like an ordinary half wave rectifier and filter circuit. Wait a minute! This entire circuit can be ready in a jiffy just by plugging into the multiple substitution box. Connect the input to a variable AC supply and you can choose the DC output voltage. If you have ever needed "one more" power supply you'll appreciate the versatility of the multiple substitution box supply. Just be certain to stay within the rating requirements of the rectifier, resistor, and capacitor.

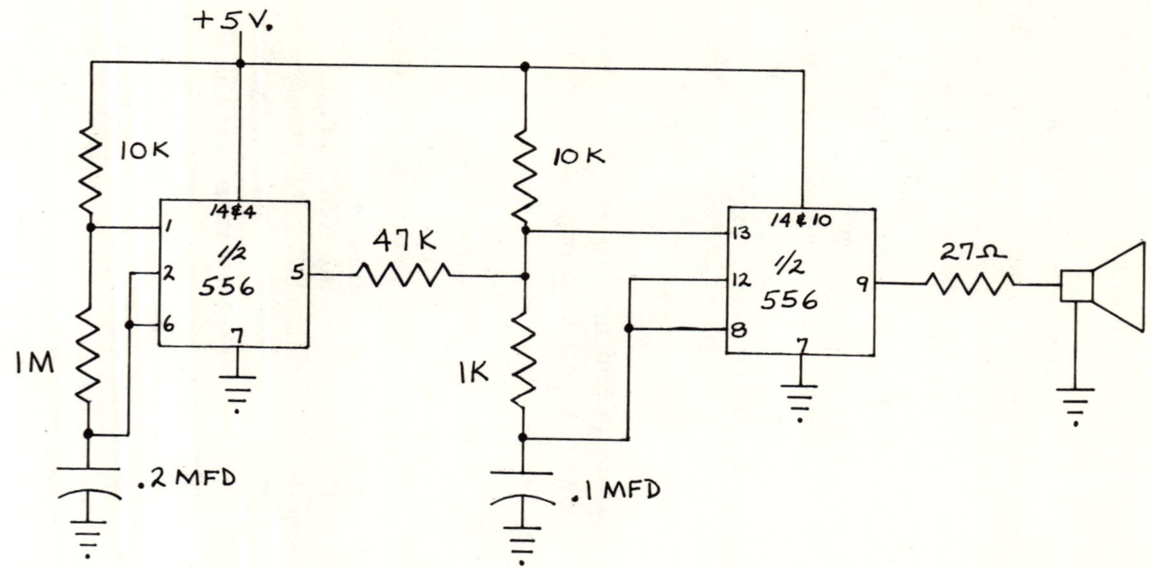

Figure 5-5B. Dual tone alarm.

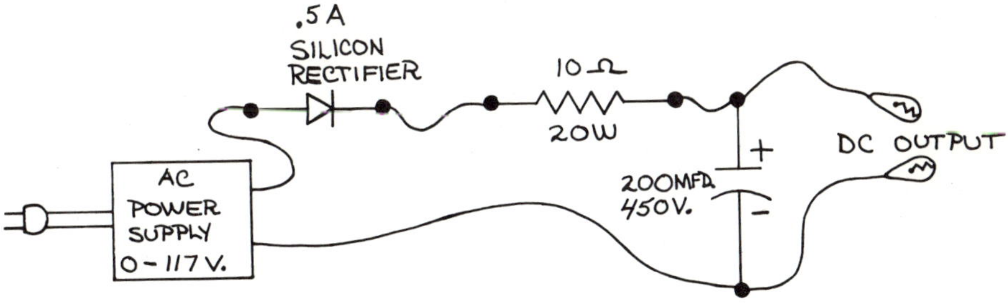

Figure 5-5C. Instant power supply circuit.

6 How to Successfully Utilize Electronic Surplus Parts

Surplus electronic parts . . . the big bargain for electronic enthusiasts. In this section you will read about ways to obtain surplus parts and you'll also find practical methods for identifying parts and values. Parts removal can be a time consuming problem. Learn some proven methods that make short work out of electronic apparatus disassembly. The last part of this section will deal with effective techniques for testing surplus components. Most electronic parts can be checked easily and speedily with minimum fuss. Read on and save yourself some money.

6-1. STEPS TO OBTAINING SURPLUS PARTS

Electronic surplus parts are the answer for many design and project applications. The majority of retail electronic parts stores do not stock or want to bother with individual parts. The few stores that do sell individual parts charge so much that many good projects are priced right out of your budget. For countless electronic construction enthusiasts surplus electronic parts mean the difference between building a project or forgetting it.

Most large cities have retail electronic surplus stores. Look in the telephone yellow pages for the address and telephone number if you're not familiar with the local surplus store. Visit the store. Be sure to give yourself enough time to browse around inspecting all the shelves, cabinets, counters, nooks and crannies. You might be surprised at the quantity and quality of components you'll run across.

You'll probably find some stores neatly organized with the parts in order and categorized. Other stores are so disorganized that you'll have to walk sideways to get into the place. In any event you will find all sorts of electronic goodies stashed away. The chances are you'll find most of the things you want. Resist the urge to impulse buy and stick with the parts you really need. However, if there is a particularly good buy that you know will eventually be needed, buy it; by the next visit it will probably be gone.

Here's an example of some prices that you might have to pay:

Item	Typical Retail Price	Typical Surplus Price
Power step-down transformer 24V CT @ 2A	$10.00	$3.00
Red LED, plastic encapsulated	.50	.15
Miniature push-button switch	1.50	.25
Electronic hardware nuts, bolts, washers, etc.	1 to 2¢ each	Sold by bag or pound
Carbon resistor 1/2W	.15	.03
Electrolytic capacitor 1,000 MFD @ 50 V	3.50	.75
Quadrac 10A @ 200 PIV	3.00	1.00

As a rule of thumb expect to pay about one-third of the retail value. Of course every so often you'll find values underpriced by much more than 30%. Part of the fun when you walk into the surplus store is that you might stumble onto the bargain of a lifetime!

Other sources for inexpensive electronic parts are the numerous mail order companies. If you are not familiar with these companies look in the last pages of any electronic magazine for their ads. You'll find most of their prices hard to beat, especially for solid state devices. Make certain you know what you are ordering and how to check them when they arrive. Some of the parts are marginal or second and may not work in all applications. The companies usually are very fair and stand by their merchandise but nevertheless it's a nuisance if you have to return too many items.

As long as you are willing to wait a week or two, parts ordering through the mail is really convenient. It saves a lot of time, gasoline, shoe leather and patience. Besides, it's kind of fun to anticipate the delivery of your package.

Surplus parts are available from some original parts manufacturers. The catch is that you have to live near a facility to benefit. Also many companies only sell to employees or friends of employees. The vast surplus warehouses of the Federal government are another source of electronic surplus. Uncle Sam gives other branches of the government first choice, then schools and non-profit organizations.

6-2. PRACTICAL METHODS FOR IDENTIFYING PARTS AND VALUES

Resistors are one of the easiest components to identify and determine value. Most resistors are marked with color bands or the value printed on the resistor body. You can generally determine wattage capability by the physical size of the resistor. Figure 6-2A shows the actual size and wattage ratings for an assortment of resistors.

Occasionally you will find precision resistors oddly marked as shown in Figure 6-2B. The last number on the right is the number of zeros to add; it is similar to the third color band on conventional color coded resistors. When in doubt about resistor value it's a simple matter to measure the resistance with an ohmmeter. Most resistors can't be measured accurately in the circuit because of other parallel paths. Disconnecting one end before measuring is the answer. Remember when measuring high value resistors to keep your fingers from touching the resistor leads or ohmmeter prod tips. Your body resistance will make all the readings erroneous.

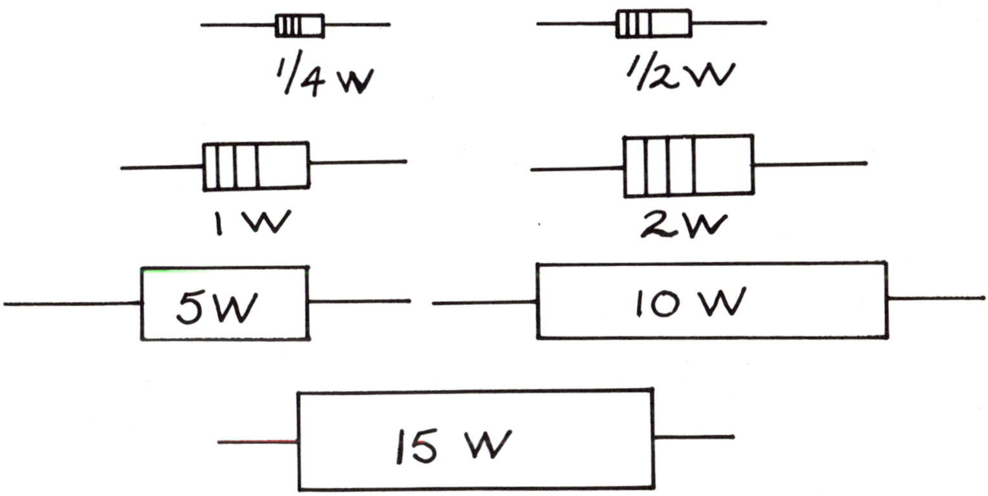

Figure 6-2A. Power rating of actual size resistors.

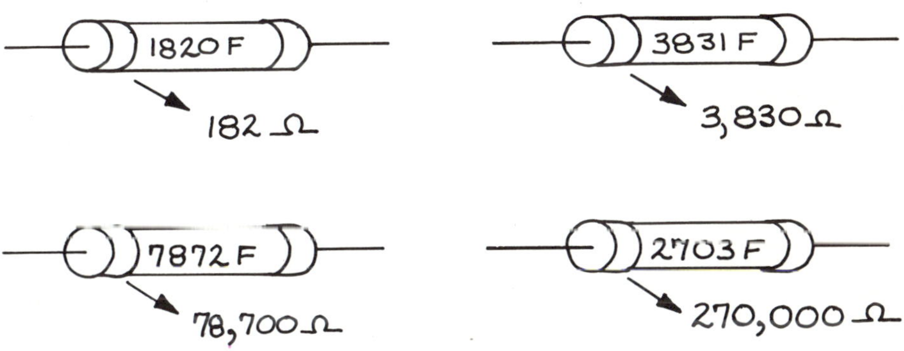

Figure 6-2B. Identifying unusual precision resistor labels.

Many potentiometers have their value stamped on the back of the case. However, many are not identified, so you'll have to make a resistance measurement between the two outside terminals to find the value. Once you find the value take time to write it on the case with a permanent felt tip marker.

Generally speaking potentiometer wattages can also be determined by physical size. The exception is when the pot is carbon or wirewound. Wirewound pots pack a lot more wattage in a much smaller space. You can usually determine a wirewound pot by slowly and delicately turning the shaft. You'll feel a slight drag as the wiper moves over the wire shoulders. A good carbon pot will turn more smoothly.

One other thing to consider when measuring pot resistance is the type of taper. Linear taper is the most common; the resistance increases uniformly as the shaft is turned. The audio taper is used for pots in audio circuits; the resistance increases in a logarithmic manner. This means that a greater percentage of the total resistance is at one end and not uniformly distributed over the entire pot range. When determining taper with an ohmmeter measure between the middle and one outside lug of the potentiometer.

Capacitors are usually well marked and easy to identify. The capacity value and working voltage are usually the most significant markings that are of interest to technicians. Figure 6-2C shows a sample of the common types of capacitors.

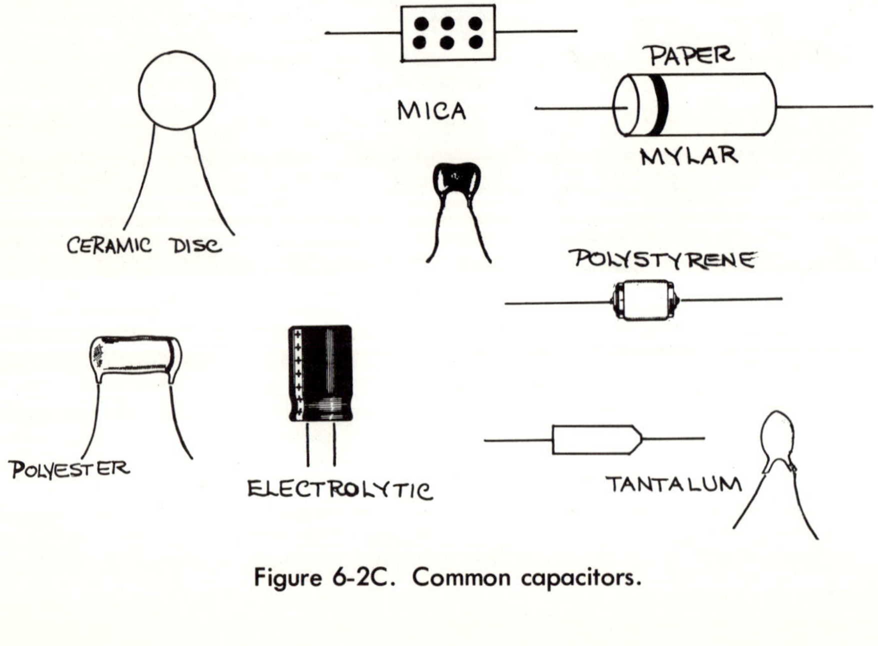

Figure 6-2C. Common capacitors.

Ceramic capacitors are usually disc types but are also made in tubular and feedthrough styles. Capacity markings such as .02, .0068, and .0015 are in microfarads. Labels in whole numbers (22, 68, 390, 680, etc.) are in picofarads. Ceramics' capacity usually range from 5 PF to .05 MFD. Working voltage as high as 1 KV is common. General purpose ceramics can be substituted for mica, paper, or plastic film capacitors in non-critical coupling, filter, and bypass applications.

Some ceramic capacitors are temperature-compensating and are marked as follows:

P 100: P means positive temperature co-efficient. For every degree of temperature rise between +25 and +85 C the capacitance will increase 100 parts per million.

N 100: N means negative temperature co-efficient. For every degree of temperature rise between +25 and +85 C the capacitance will decrease 100 parts per million.

N P O: No capacitance change between temperatures +25 and +85 C.

Modern mica capacitors are epoxy dipped and ink marked for easy identification. Older mica capacitors are read as shown in Figure 6-2D. Mica capacitors make good substitutes for ceramic capacitors if temperature compensation is not a factor.

Electrolytic capacitors are generally well marked with capacity in microfarads and working voltage labels on the case. They are also marked with polarity. Be sure to identify the positive and negative terminals before soldering into a circuit. Polarity labels are usually shown with + and − labels next to the appropriate lead. In metal case electrolytics the case is usually assumed to be the negative terminal. When replacing electrolytics in filter circuits it's almost always permissible to increase the capacity and voltage values if the exact part is not available.

IC's are becoming more and more abundant on the surplus market. The hassle is to determine the relationship between the number stamped on the case and what's inside. If the number is used throughout the industry the problem is not too bad. Any complete electronic parts catalog will probably list the IC and

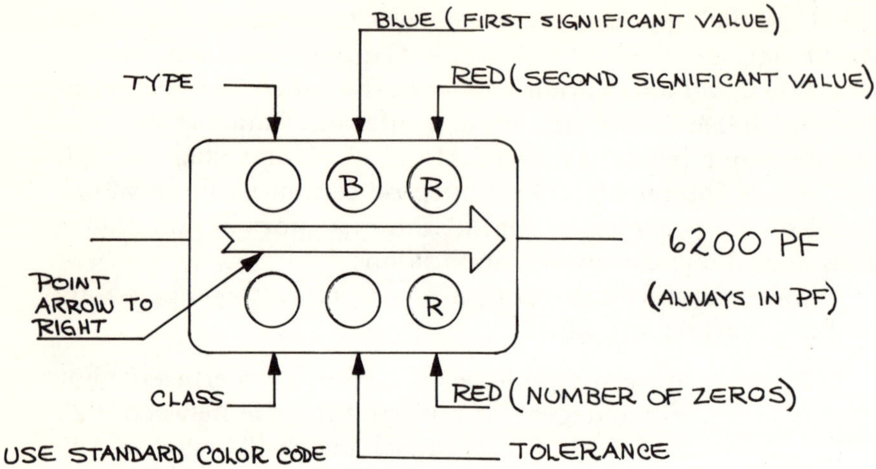

Figure 6-2D. Identifying a mica capacitor's value.

give a brief description of its function. Finding pin numbers and other pertinent information may be a bit more difficult. All IC manufacturers have catalogs describing their products in detail. These catalogs are relatively inexpensive and a must for finding the information needed. Obtain these helpful catalogs through the mail by ordering directly from the manufacturer. The following list gives the name and address of some IC manufacturers.

Advanced Micro Devices
901 Thompson Place
Sunnyvale, CA 94086

Fairchild Semiconductor
313 Fairchild Dr.
Mountain View, CA 94040

General Electric
Semiconductor Prod. Dept.
Electronic Park
Syracuse, NY 13201

ITT Semiconductors
3301 Electronics Way
West Palm Beach, FL 33407

National Semiconductor
2900 Semiconductor Dr.
Santa Clara, CA 95051

Raytheon Semiconductor
350 Ellis St.
Mountain View, CA 94040

RCA
Electronic Components
Harrison, NJ 07029

Signetics
811 E. Arques Ave.
Sunnyvale, CA 94086

Motorola Semiconductor
Box 20912
Phoenix, AZ 85036

Stewart Warner Corp.
730 E. Evelyn Ave.
Sunnyvale, CA 94086

Sprague Electric Co.
481 Marshall St.
North Adams, MA 01247

Texas Instruments
Box 5012
Dallas, TX 75222

For IC's with in-house numbers you'll have to get the information directly from the manufacturer providing you know who made the device. Most IC's are identified with the manufacturers symbol, initials, or abbreviation.

There is an IC master catalog available that surveys the entire IC industry. A list of topics included in this book is shown below:

1. Parameter selection guide
2. Alternate source lists
3. Military device directory
4. Manufacturer's data
5. Application notes
6. Manufacturer's directory

Although this book is fairly costly it certainly is a time saver and well worth the money. Just a handful of identified salvaged IC's will more than pay for the book. Figure 6-2E shows the most comprehensive IC catalog on the market. It's called the IC Update Master and is available from IC Update, 1287 Lawrence Station, Sunnyvale, CA 94086.

6-3. PARTS REMOVAL—NEAT AND QUICK

Very often surplus electronic equipment and subassemblies can be purchased in bulk. The individual parts obtainable in this fashion usually cost very little. The chances are you'll buy bulk electronics assemblies to obtain a few wanted items. The initial cost pays for these items and everything left over can be considered a bonus. Most parts obtained in this manner are good. Many times the equipment is brand new or has very few hours of actual service. In this age of rapid obsolescence electronic apparatus becomes outdated many years before its expected life span.

Courtesy of IC Update, 1287 Lawrence Sta., Sunnyvale, CA 94086

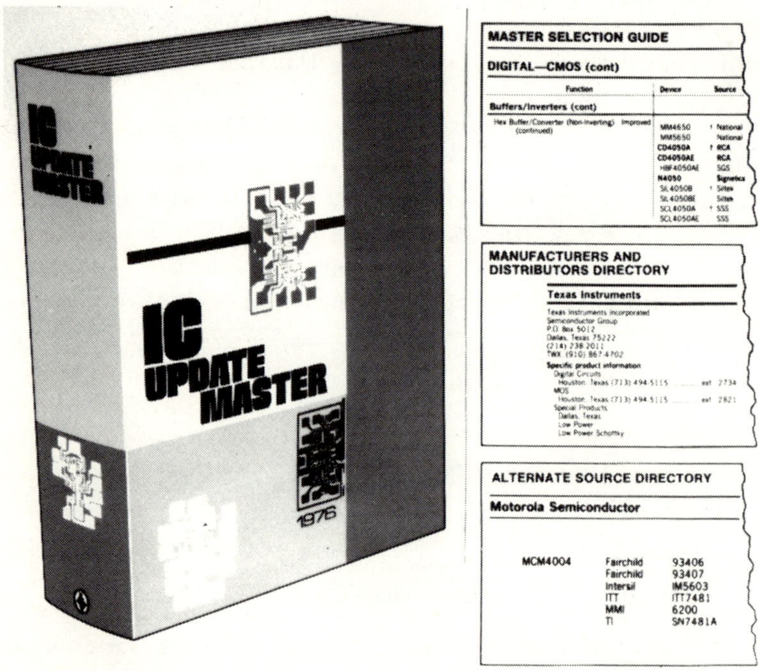

Figure 6-2E. Comprehensive IC catalog/directory.

Disassembly and removal of parts is the biggest drawback when purchasing bulk equipment. It takes time but you may be surprised how fast disassembly can be if you follow some time saving principles. Tackle the job during some free time when nothing else is pressing. It's a good way to relax while listening to the ball game.

The first step is to take apart the equipment mechanically. Usually an assortment of screwdrivers, nutdrivers, and small open end wrenches are all that's needed. Reassemble all the screws, washers, lockwashers, nuts as they are removed. This will save a lot of time later when sorting the hardware. Don't bother unsoldering any wire connections at this stage since it's far too time consuming. Simply snip the wires with a heavy duty pair of dykes, see Section 4-1. This fast, efficient method will not only save time but prevent many parts from being broken accidentally. Leave an inch or two of wire on each connection so wire removal will be easier at a later date. Once in a while you'll run into parts riveted in place. Don't let this stop you. Just center punch the rivet head and drill out with a slightly larger drill bit.

Keep related items together during disassembly. A good storage system, see Section 2, will provide a place for items not needed immediately. The best time to desolder and remove wire connections from parts is just before they are going to be used. There are going to be a lot of parts on stand-by for years and some that will never be used so why waste your time removing the wires?

You'll probably be amazed at the number of useful hardware items obtainable on bulk surplus apparatus. Screws, washers, nuts, spacers, brackets, insulators, grommets, feet, terminal strips, posts, jacks, are just a few of the items that you'll discover.

The next big job after mechanical disassembly is individual parts removal. Most of the parts probably will be soldered in a PC board. The trick is to remove them fast without stressing or breaking off the leads. For many PC parts the quickest way is to use dykes to snip them off even with the board. About 1/8 inch of lead will be wasted but the part will not be damaged. Even though used PC parts have short leads they certainly can be used again especially if you do a lot of experimenting or project building. Longer leads can be spliced and soldered on in a jiffy; covered with spaghetti they will be like new.

Parts that are mounted directly against the PC board will have to be desoldered for removal. Section 4-4 describes some successful desoldering techniques. Figure 6-3A shows a new tool that is especially built for extracting 14 and 16 pin DIP IC's. This tool grips and heats the IC pins from the part side of the PC board. As soon as the solder melts the IC can be pulled out of the board even if the pins are bent against the copper foil. Once the IC extractor reaches operating temperature, about five minutes, the tool is ready to go; IC's are out in a matter of seconds.

Courtesy of Wall-Lenk Mfg. Co.

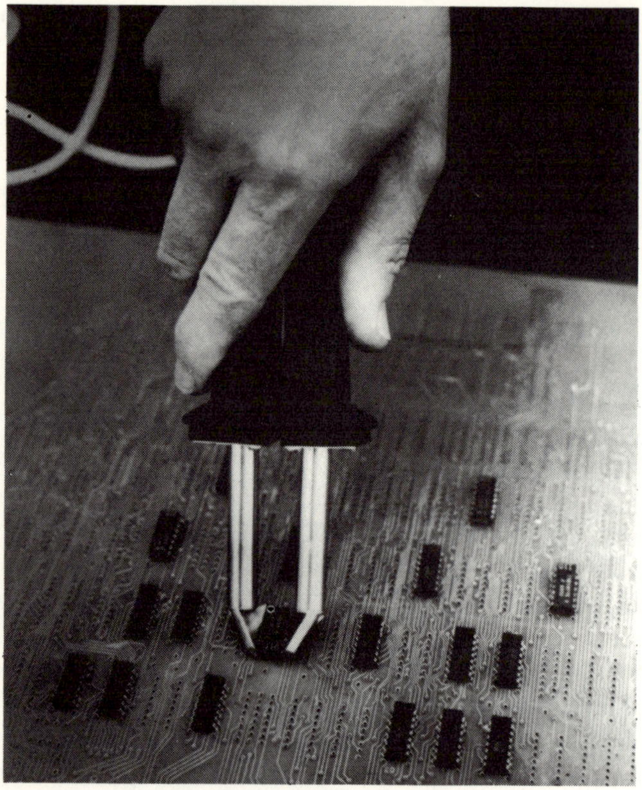

Figure 6-3A. IC extractor.

6-4. EFFECTIVE METHODS TO TEST SURPLUS COMPONENTS

It's a good policy to check all surplus parts before soldering them into the circuit. The majority of the components will be good but you can't be sure. Resistors are easy to check with a VTVM or DVM ohmmeter. Checking them now may save a lot of headaches later. A capacitor checker will take the guess work out of identifying bad or out-of-tolerance capacitors, see Figure 6-4A. Use this same instrument for measuring precise resistances and finding unknown inductances of coils and chokes.

Refer to Section 5-4 before plugging in that new looking tube. Transistors and diodes can be easily checked using the transistor-diode checker described in Section 8-2. You might need to check a few unijunction transistors as well; Section 8-4 has the answer. How about the thyristor family . . . SCR's and triacs. Section 8-3 shows a tester circuit especially made to test these solid state devices.

Courtesy of Heath Co.

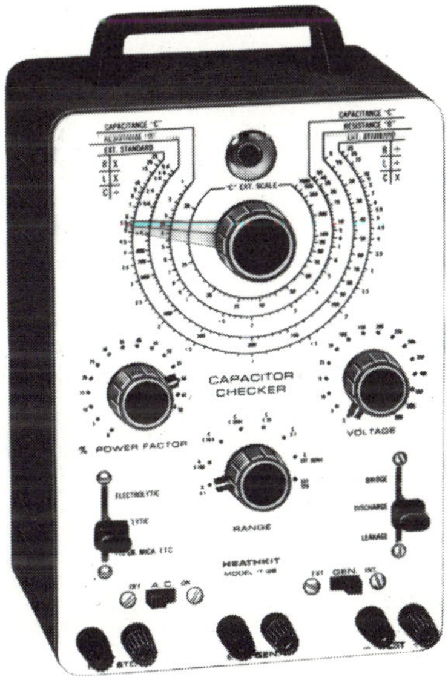

Figure 6-4A. Capacitor checker.

Zener diodes can be checked as rectifiers. Measure the forward and back resistance, refer to Section 5-1. If the zener checks okay continue to test it using the circuit shown in Figure 6-4B. Choose the load resistor so that approximately one-half of the maximum zener current flows through it. Apply power to the circuit. VTVM A will monitor power supply voltage and VTVM B monitors zener voltage. Meter B will follow meter A as power supply voltage is increased until the zener point. When this breakdown point is reached meter B will remain at the zener voltage while VTVM A will continue to read the power supply output voltage.

Here's an example. Suppose you want to check a 20V 1W zener diode. Use the following formula to find R_L:

$$R = \frac{E^2}{P}$$

$$R = \frac{20^2}{1}$$

$$R = \frac{400}{1}$$

$$R = 400 \text{ ohms}$$

Double the resistance so that only one-half of the maximum current flows. An 800 ohm resistor will do the job. Find the resistor wattage as follows:

$$I_{zener} = \frac{E}{R} \qquad P = I^2R$$

$$P = .05^2 \times 800$$

$$I_{zener} = \frac{20}{400} \qquad P = .0025 \times 800$$

$$I_{zener} = .05 \text{ A} \qquad P = 2\text{W}$$

Check any voltage zener using this technique.

Before using any mechanical switching device such as relays and switches be sure to check their contacts for oxidation. Often these devices have been gathering dust for a number of years and a film of oxidation will develop on all the contact surfaces. If you can measure any resistance with an ohmmeter across a closed contact you can bet it will cause trouble. The solution is generally quite simple. Shoot the contacts with aerosol tuner or contact cleaner. If the contacts are inaccessible squirt the cleaner into any opening, down the shaft or button. Operate the device manually 25 or more times. You should now be able to measure zero ohms across the closed contacts. This procedure will take care of the majority of contact problems.

For contacts that won't clean with this procedure try passing a clean white business card through the closed contacts a number of times. For more stubborn cases try cleaning the contacts with a typewriter eraser. Filing is the last resort since it removes precious contact surface metal.

We have left the toughest component of all to check until

last . . . the integrated circuit. Most IC's are rough to check since there are so many varieties. A universal IC checker would be just about impossible to build. There are a few IC's that can be tested with an ohmmeter reasonably well. However most of them are just too complex for any conclusive ohmmeter test. The best and most practical way to check any IC is to build a working circuit using that particular IC in a test socket. Use a known good IC to get the circuit operating properly. Then unplug the good IC and plug in the unknown quality IC of the same type. Monitor the circuit with your test instruments. If the circuit continues to function properly the unknown quality IC must be good. Once you get set up a lot of identical integrated circuits can be tested in short order with this technique.

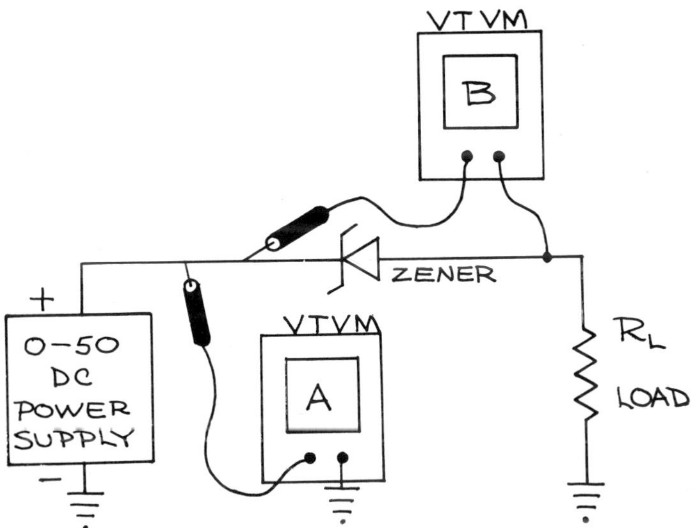

Figure 6-4B. Zener diode testing circuit.

7 Guidelines to Selecting, Making, and Using Probes and Test Leads

This section is devoted entirely to probes and test leads that are mighty useful in the electronic shop.

You'll read about the hows and whys of the oscilloscope low capacitance probe. Also find out how a demodulator probe can make short work of locating defects in radio frequency amplifier stages. High voltage measurements both DC and AC can easily be made if the correct probe is used. You can measure 25,000 volts or more as routinely as checking a flashlight battery just by using the correct tools and techniques.

Alligator-minigator-jumper leads, test clips, probes, you name it; some of the handiest and most worksaving devices are an assortment of quick-fastening test leads. Discussion of test leads is included in this section as well as neat and nifty methods for storage and organization.

7-1. BECOME FAMILIAR WITH THE LOW CAPACITANCE PROBE

A low capacitance probe, Figure 7-1A, is needed for oscilloscope waveform measurements whenever the signal is affected by the relatively high input resistance and capacitance of the scope, as well as the capacitance of the coaxial input lead. All high frequency signals are distorted and sometimes lost when using a direct type probe. Two effects occur; the signal is bypassed to ground by the scope input and cable capacitance, and/or the circuit being scoped may be loaded or detuned.

Courtesy of Heath Co.

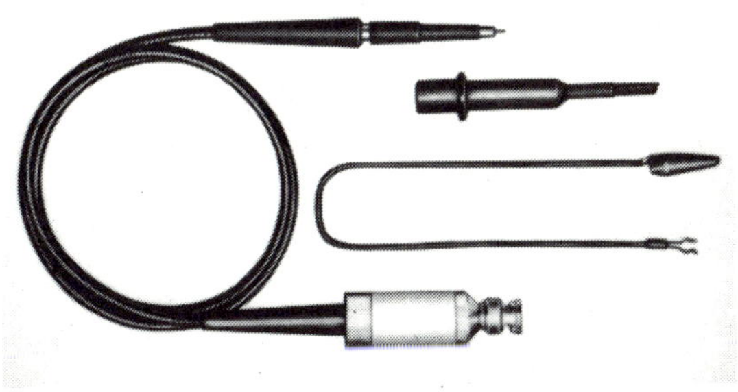

Figure 7-1A. Low capacitance probe.

Figure 7-1B shows the schematic of a typical direct scope probe and vertical input impedance network. Shielding is necessary to prevent pickup of any interfering signals. The signal will "see" a 1 meg ohm resistor and an 80 PF capacitor. The shielded cable has a capacitance of about 60 PF and the scope input capacitance is 20 PF so the combined capacitance will be 80 PF. A direct probe is useful in circuits where 80 PF will not shunt an appreciable amount of signal to ground and where 1 meg ohm will not load a circuit; for all practical purposes low impedance audio frequency circuits.

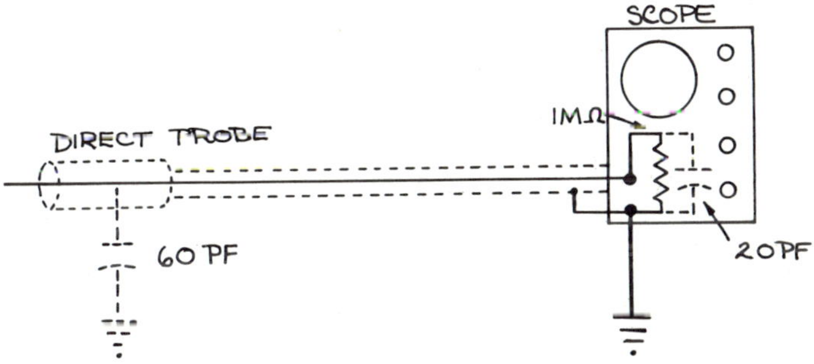

Figure 7-1B. Direct probe and oscilloscope input impedance network.

The low capacity probe, LCP, corrects the deficiencies of the cable capacitance and scope input impedance characteristics, see Figure 7-1C. Trimmer capacitor C1 in the probe is one part of a capacitive voltage divider. The cable capacitance as well as the scope input capacitance form the other part of the voltage divider. This will significantly decrease the capacitance loading effect. R_1 is built into the probe to provide higher input resistance and linear voltage division regardless of the input frequency. Some probes have a built in switch to enable the user to select either a direct or low capacity input.

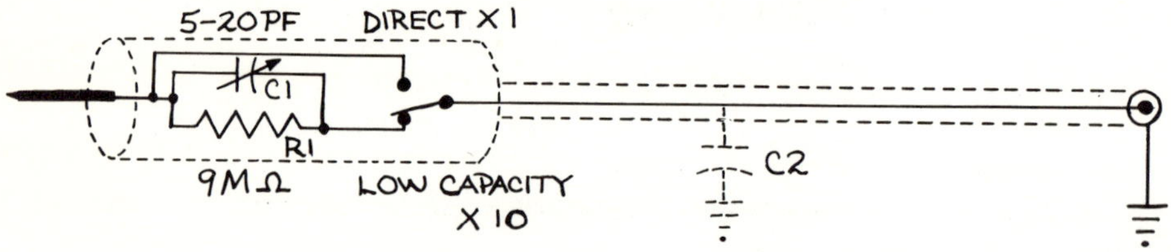

Figure 7-1C. Low capacity probe schematic.

The one drawback to the low capacity probe is that the signal is reduced 10:1. In other words if the signal being scoped has a peak to peak voltage of 10V, only 1V will be available to the scope. The parallel combination, C_1 and R_1, drops 9V. This network is designed to voltage divide 10:1 for easy scope calibration. Simply multiply the scope P to P voltage by 10 whenever using the low capacity probe. The voltage attenuation is not critical since most scopes have more than enough vertical gain to compensate for the probe loss.

Here's how to match the LCP to the oscilloscope. Observe a 1 KHZ square wave on the scope using the LCP. The square wave leading and trailing edges should be vertical and the tops and bottoms horizontal. The LCP will probably distort the square wave until matched with the probe trimmer adjustment. Figure 7-1D waveforms 1 and 2 show how the waveform may appear before the trimmer is properly adjusted. Simply turn the trimmer capacitor while watching the scope waveform. When the square wave appears as shown in Figure 7-1D waveform 3, the LCP is correctly matched to the scope. Now you are ready to make accurate waveform analysis in both high frequency and high resistance circuits with negligible probe and scope distortion.

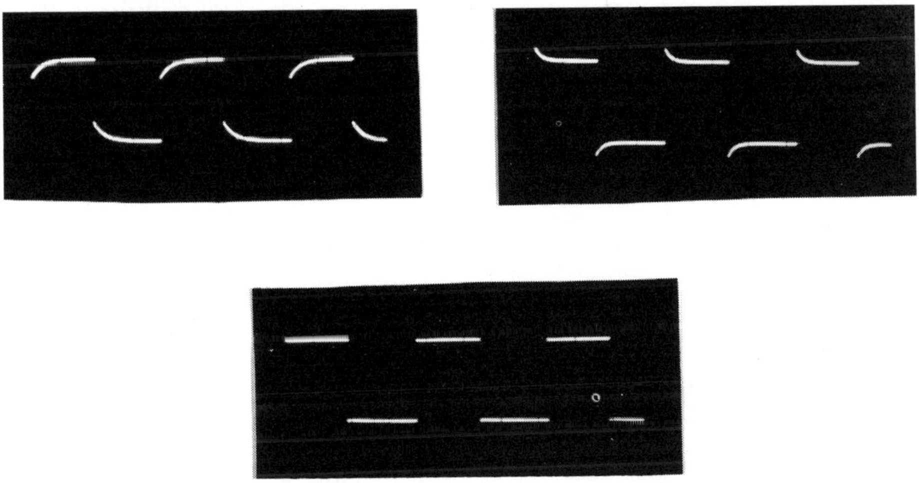

Figure 7-1D. LCP calibration using a 1,000 HZ square wave.

7-2. THE DEMODULATOR PROBE: DIAGNOSE CIRCUIT TROUBLES

The oscilloscope demodulator probe, Figure 7-2A, is used whenever modulation envelopes of radio frequency carriers are being observed. The probe effectively removes the carrier and either the positive or negative envelope, enabling the technician to signal trace RF and IF stages.

Courtesy of Heath Co.

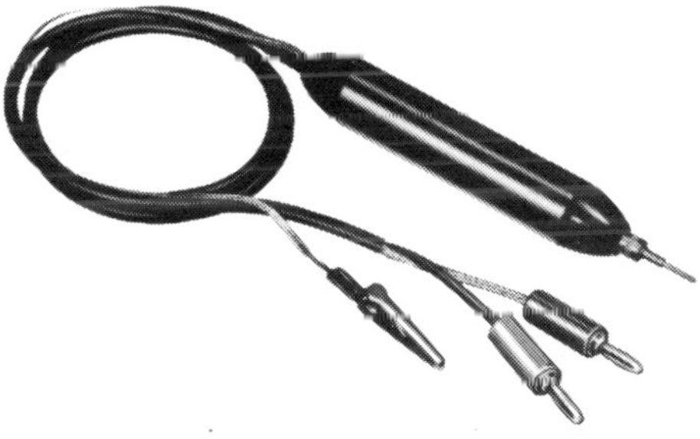

Figure 7-2A. Oscilloscope demodulator probe.

Figure 7-2B shows the schematic of a typical demodulator probe. The input capacitor (200 PF) is generally rated for about 600 volts DC so don't exceed that value. The probe housing is made of metal and grounded through the coaxial cable to prevent pickup and hand capacity interference.

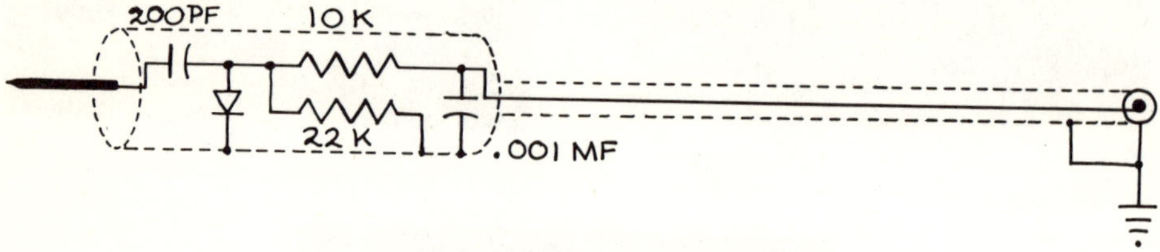

Figure 7-2B. Demodulator probe schematic.

Suppose you are troubleshooting a TV receiver and think the trouble is in the video IF amplifier chain. The symptom is good raster, no picture, and no sound. You have scoped the output of the video detector looking for the composite video signal but to no avail. Here's the perfect place for the demodulator probe. Replace the existing scope probe with the demodulator probe. Now, it's possible to continue signal tracing back into the IF's using your scope.

Figure 7-2C shows a partial schematic of a transistor IF section being scope traced. First check the primary of the 3rd IF transformer as shown. Note that scope A is placed across the primary winding; probe to the top, ground to the bottom. This is an important technique when scope tracing an IF stage. Most scope measurements are made in relation to ground, that is scope ground on circuit ground. By placing the scope leads directly across the IF transformers or coils, power supply ripple will be minimized and signal will be the predominant waveform. Note: to isolate TV and scope grounds from each other use an isolation transformer to power the TV. Keep in mind when checking IF amplifiers' output wafeforms that the scope ground (case) will assume the same potential as the collector supply.

If the composite video is not observed on scope A, next check scope B position. A video waveform here but not on scope A would indicate a defect in the third IF section. How-

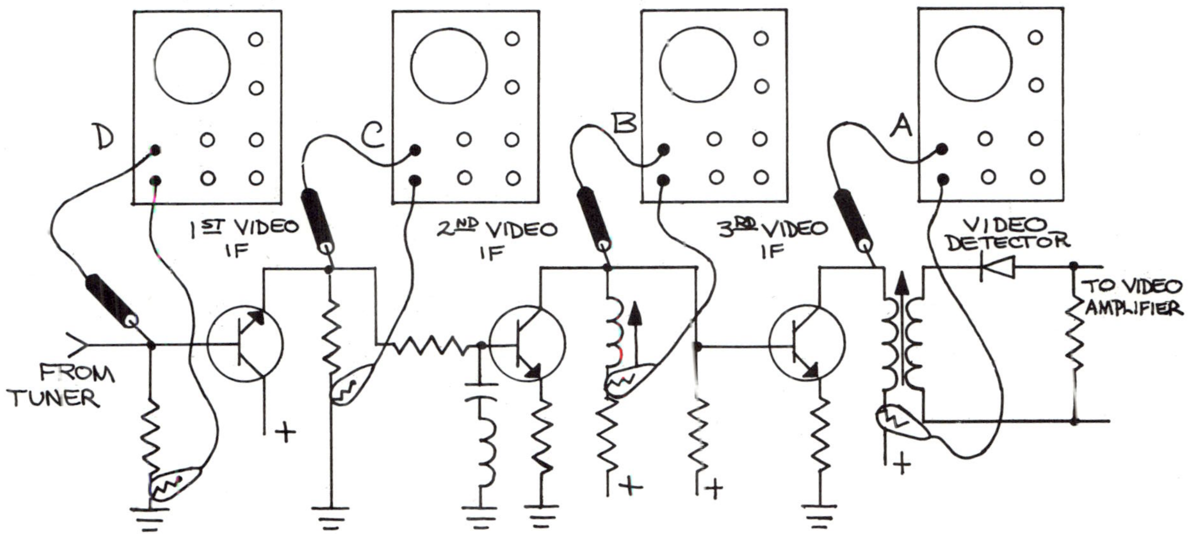

Figure 7-2C. Signal tracing an IF amplifier section using a demodulator probe and oscilloscope.

ever if scope B still shows no video move the probes to scope C position. A good waveform would indicate a bad second video IF. No waveform means scoping back one more stage, scope D.

Remember when scope tracing IF stages that the signal is very weak especially as you move closer to the tuner. (Tuner output may only be .5 MV.) The scope vertical attenuator controls will probably have to be set at their most sensitive settings. As each IF amplifier does its job the video signal's amplitude will increase. By the time the signal is detected by the video detector the peak voltage should be about 1 to 3 volts.

7-3. ALL ABOUT HIGH VOLTAGE PROBES

Caution: high voltages are extremely dangerous; a familiar warning to anyone who has worked with high voltages. One of the devices that allows safe high voltage measurements is the high voltage probe, see Figure 7-3A.

Courtesy of Heath Co.

Figure 7-3A. High voltage probe.

Most HV probes are designed to measure DC volts and operate in conjunction with a VTVM or DVM. Figure 7-3B shows the schematic of a typical VTVM HV probe. Since a VTVM input resistance is 11 megohms and the probe resistor is 1089 megs, the combined VTVM and probe resistance is 1100 megohms or 100 times as great as the VTVM without the probe.

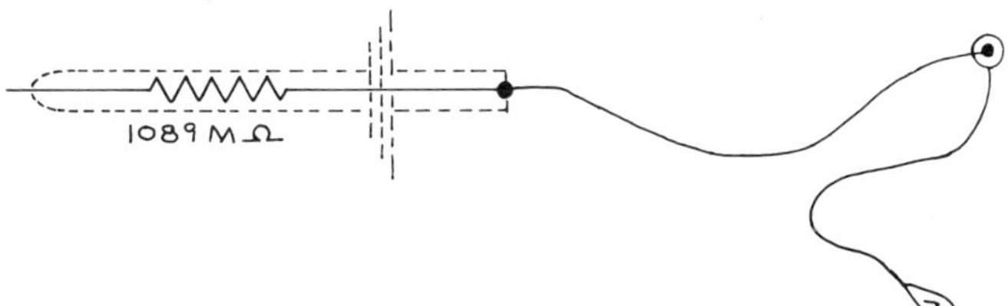

Figure 7-3B. VTVM HV probe schematic.

When using the probe, simply multiply the voltage range by 100. For instance a 100 volt range becomes a 10,000 volt range, a 300 volt range becomes a 30,000 volt range. The highest range on many VTVM's is 1,500 volts. It seems like multiplying by 100 would give a 150,000 volt range. Stop: don't try to measure voltage in excess of 30,000 volts because that's the safe limit for many probes.

Since the HV probe increases the meter input resistance to 1,100 megohms, critical measurements in very high resistance circuits can be made without loading. A 1.5 volt range becomes 150 volt range with extremely high input impedance. Keep this in mind the next time you have to make an accurate reading in a high resistance circuit. It's often hard to remember that an HV probe can be and should be used in low voltage-high resistance circuits.

Figure 7-3C shows an HV probe which features a built-in meter in the handle. This probe is really handy and most convenient to use. Simply connect the ground lead to the chassis

Courtesy of ITT Pomona Electronics

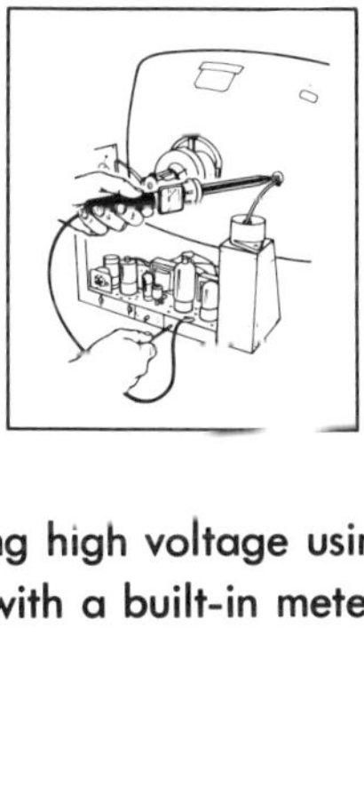

Figure 7-3C. Checking high voltage using a DC high voltage probe with a built-in meter.

and measure high voltage with the probe; read HV directly from the built-in meter. Very simple—no interchanging meter probes, no warm-up.

Before using the HV probe read these precautions:

1. Don't measure in excess of 30,000 VDC unless the probe is specifically made for higher voltages.
2. Always clip the ground lead on first before measuring high voltage.
3. Connect the probe to the VTVM or DVM before contacting an HV point.
4. When possible connect HV probe to circuit before power is applied. Remove after power is off.
5. Keep your hand behind the circular handle shield and your other hand in your pocket.

A 10 KV AC high voltage probe is shown in Figure 7-3D. Use this probe to measure high AC voltages such as you would find in neon sign transformers or other high AC voltage apparatus. The unit shown has a 2,000 ohms/volt sensitivity and is ± 3% accurate for frequencies from 30 HZ to 1,000 HZ.

Courtesy of ITT Pomona Electronics

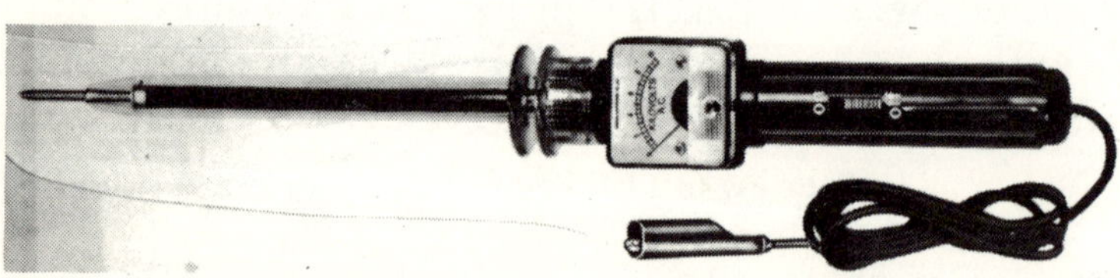

Figure 7-3D. 10 KV AC high voltage test probe with built-in meter.

7-4. KINKLESS TIPS FOR TEST LEADS

Test leads can often turn a simple repair into a nightmare or vice versa. If you have ever used an open clip or test lead to make a jumper, you know what type of wild goose chase can develop. On the other hand a reliable assortment of test leads can often do wonders turning seemingly tough dogs into easy repairs.

Let's start with simple 12″ clip leads; an alligator or minigator clip on each end. Don't throw these leads together in a hurry with any old wire or buy the cheap ready-made kind. Good materials and painstaking construction will pay off in the long run. Use test prod wire. It's expensive but will hold up to the kind of usage a clip lead encounters. Generally 20 gauge wire with 42-36 gauge strands is ideal. The insulation is usually rated for 10,000 volts or more. Use a good quality clip. There are cheap clips on the market that aren't worth the powder to blow up. They don't grab properly, they slip, and they are hard to operate. Make enough clip leads in both variety and size so you'll always have enough. There's nothing more frustrating than needing a clip lead when they are all being used.

Always use vinyl insulators on each clip; you'll never have to worry about the clip shorting out a circuit. Be super careful when stripping the insulation so no strands are nicked or cut, refer to Section 4-2. When soldering the lead to the clip don't permit the solder to flow back into the strands. This phenomena called wicking draws solder by capillary action underneath the lead insulation. As a result the wire becomes stiff at the point where maximum flexibility is desired. Conductor heat sinks, anti-wicking tools, are specially made to stop wicking, see Figure 7-4A. However any spring-loaded heat sink or

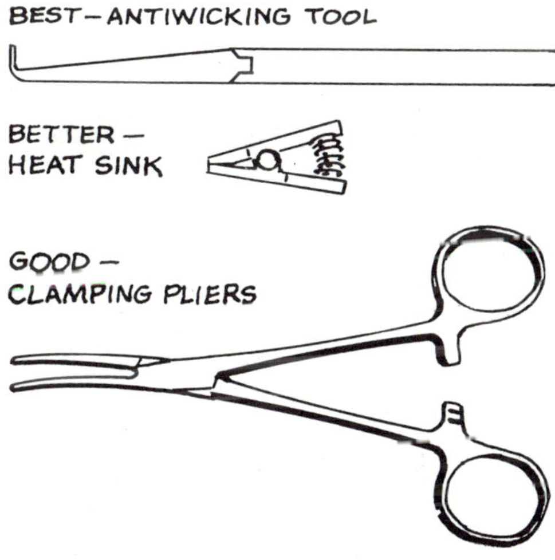

Figure 7-4A. Conductor heat sinks.

clamping needle nose pliers will reduce wicking. Connect the conductor heat sink between the clip lead and the insulation on the test lead wire before soldering.

If you have ever had a test lead pull out of an instrument when making a critical measurement you'll appreciate the following suggestions. Make it very difficult for a test lead to be accidentally pulled out of its jack. Use right-angle plugs for all your banana and tip jacks, see Figure 7-4B. Another pull-out resistant plug is the double banana or pin tip plug. See Figure 7-4C. In addition to staying in their jacks, these plugs permit stacking for multiple connections. Just be sure that the jack centers can accomodate the double plug. The standard double plugs are constructed on 3/4″ centers; the miniature plugs on 1/2″ centers.

Figure 7-4B. Right-angle instrument plug.

Courtesy of ITT Pomona Electronics

Figure 7-4C. Double plug.

As long as you are going to make your instrument leads pull-out-proof why not splurge and make some of them retractable? These cords are generally 5′ long but coil up to less than a foot. How nice to have instrument test leads that coil themselves up out of the way when not being used!

Make sure to have a couple of test prods with needle tips. These tips are strong and sharp, designed to penetrate easily a flux residue on a solder joint. Another good trick using these prods (although not always recommended) is piercing wire insulation to make a measurement. When both ends of a wire are inaccessible you can't beat this technique. The needle tip will leave a very small hole in the insulation that usually closes when the prod is removed. Don't attempt this technique on high voltage leads because arcing may occur. Along these same lines there are test clips available similar to large alligator clips that have an insulation piercing needle built into the inside of the jaws. It's just a matter of clipping it around an insulated wire, then twisting and pressing slightly to enable the insulation to become punctured.

Figure 7-4D shows a pair of test leads with 3″ steel tips on the prod. These beauties are perfect for getting into tight places to make that "impossible" measurement. Don't worry about shorting out those long tips because they are covered with vinyl insulation.

Figure 7-4D. Extra long test lead prod tips.

In this age of miniaturization don't overlook one of the most convenient and handiest test clips of all. It's a mini-test clip called the grabber, see Figure 7-4E. The plunger type action clips are completely insulated except for the connection point. It's a snap to make a positive connection deep into a maze of IC's. Once the hookup is made the connection remains tight and won't pop off. They are also designed to slide down over square wire-wrap pins. It's easy to attach the test lead to the clip. Simply pull off the thumb cap to expose the connection. Next thread the wire through the hole in the thumb cap. Strip the wire and slip it into the contact hole, solder, and slide the thumb cap back on, finishing the job. You'll probably find yourself replacing a lot of alligator and minigator clips on your test leads with these grabbers.

Courtesy of ITT Pomona Electronics

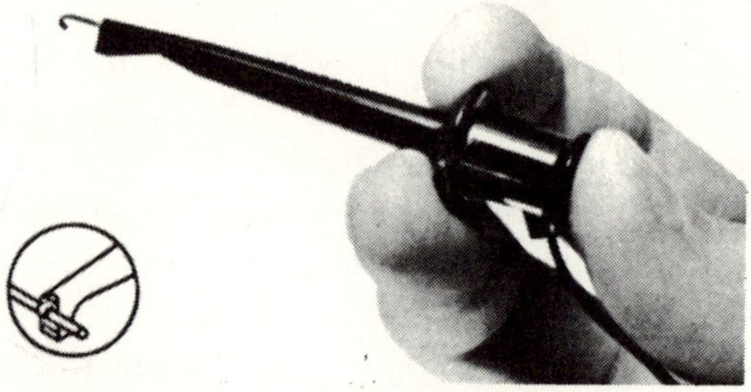

Figure 7-4E. Mini-test clip-grabber.

7-5. TECHNIQUES FOR PROBE AND TEST LEAD STORAGE

Pick a place for probe and test lead storage that's easily accessible. These items are continually used in the electronic workshop so position their storage facilities in an area that will cause you no strain. Racks, lead holders, hooks, or what ever, can be fastened in otherwise wasted space. For instance the inside of a cabinet door or the side of your workbench make ideal storage areas. You might be surprised to find out how many test leads and probes can be stored in a relatively tiny area.

You can make a holder for probes from wood or metal depending on your preference. Figure 7-5A shows a wooden probe holder that's easy to make and use. This particular holder was made from scrap wood in a jiffy. Make the bottom section as long as necessary (X) to fit your space requirements. The slot width (Y) will be determined by your probe or cable diameter; 1/4″ and 1/2″ slots will fit most applications.

After the slot positions are determined, saw along the layout lines. One whack with a chisel the width of the slot will neatly remove the slot wood. Nail the back piece to the completed bottom section and fasten to a wall, cabinet, workbench, etc. If you don't have room for the leads to dangle, screw in a cup hook at the end of each slot. Loop the cable a couple of times around the hook for compact storage.

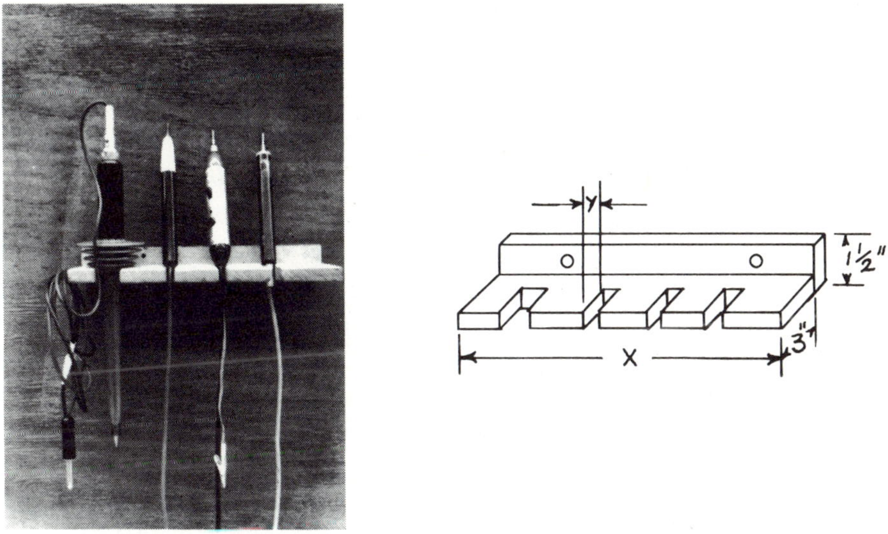

Figure 7-5A. Wooden probe holder.

Figure 7-5B shows a ready made test lead holder that can hold a multitude of cables and clip leads. Each slot can hold a

Courtesy of ITT Pomona Electronics

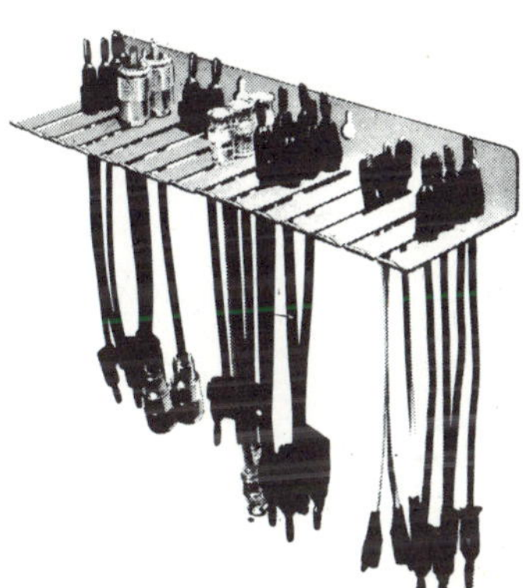

Figure 7-5B. Test lead holder.

number of identical leads making it a snap to keep leads separated but easy to store and remove.

Storing clip leads can be accomplished with the simplest items. A long nail similar to one used for fastening rain gutters can hold a zillion clip leads. Simply clip each lead onto the nail. Another quick and easy clip lead hanger is nothing but a length of #14 gauge solid house wire connected horizontally by two screws or nails, see Figure 7-5C. Even non-clip lead type test leads (banana, tip, phono, BNC, etc.) can be hung from the wire for convenient storage.

Figure 7-5C. House wire clip lead hanger.

Need a storage rack for odd-ball leads—the kind that never seem to fit or stay on lead holders? Figure 7-5D shows a spring clothes pin rack that can't be beat. It's inexpensive, easy to build, and really works. Fasten the clothes pins to the wooden rack body with nails driven through the opening in the clothes pin coil spring. Leave enough space between each clothes pin for finger space. The clothes pins grab tightly, hanging on to most leads, plugs, and adapters, with more than enough pressure. Build it; you'll like it.

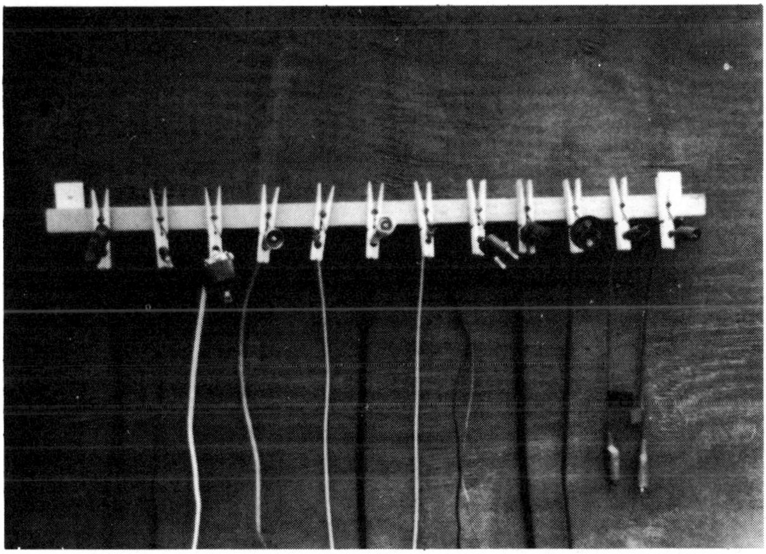

Figure 7-5D. Clothes pin lead rack.

Why don't all test instrument manufacturers include a place on their instruments for lead, prod, and probe storage? Most instruments are equipped with leads of some kind. What do you do with them when you keep them with the instruments? There are usually two choices. Let the leads hang tangling and cluttering up the work area or coil them and store under the handle. There's a better solution—make a test lead storage bracket for each of your instruments. They can be made out of scrap sheet aluminum and often fastened to the case with an existing screw or two. Figure 7-5E shows a few instruments with custom built lead and probe storage brackets. It's well worth the time for this added convenience.

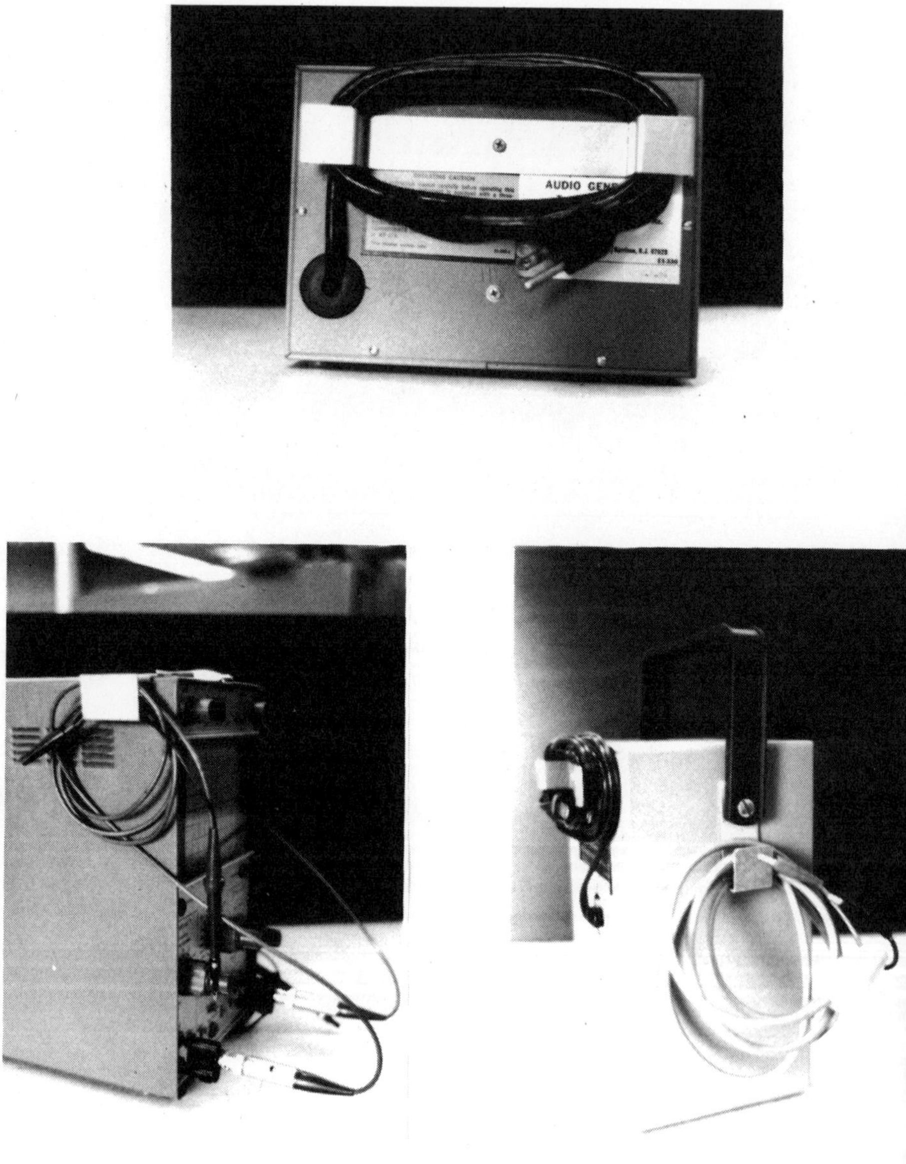

Figure 7-5E. Custom built lead and probe storage brackets.

8 Plans for Easy-to-Make Test Equipment

This section shows schematics and details operation of six worthwhile test instruments. Each one of these projects is necessary for the complete electronic workshop.

Build a transistorized signal injector capable of broadcasting an airborne radio frequency signal as well as injecting intermediate frequency and audio signals. Find out how to construct an LED readout transistor-diode checker. Multicolored LED's indicate transistor and diode performance parameters. Need a reliable means to test thyristors? SCR's and triacs can be analyzed in a jiffy using the plans shown in Section 8-3.

Wire together the unijunction transistor tester as shown in this section and it will actually chirp when good unijunctions are tested; for defective unijunctions it will be as quiet as a mouse. Section 8-5 describes a variable low voltage, high current semiregulated DC power supply that's really a beauty. In the last section you'll find plans for a triac controlled AC power supply that can be used for everything from a lamp dimmer to a motor speed controller.

8-1. THE SIGNAL INJECTOR: A POCKET LABORATORY

Here's a little instrument that should be a part of every electronic workshop. A signal injector is simply a small audio multivibrator that produces a fixed frequency distorted square wave. Although the tone produced is audio, the harmonics of the square wave can be in the IF and RF range. So when you click the switch "on" you have a miniature hand-held audio frequency generator and radio frequency transmitter that's ready to go.

Figure 8-1A shows the schematic of a typical signal injector. This is one project where all the parts can be found in just about every electronic workshop. Most technicians like to build their injector as small as possible to make it more convenient to use and store. Printed circuit construction is ideal but not absolutely necessary. The layout is not critical and will work with any wiring technique.

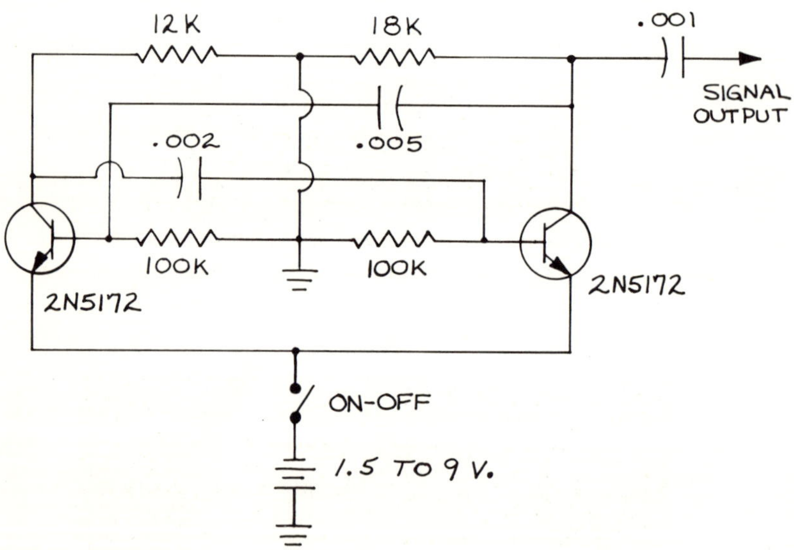

Figure 8-1A. Signal injector schematic.

Build the injector into a probe case, refer to Figure 8-1B. The PC layout is actual size and will fit easily into many probes. Choose a miniature type toggle, push button, or slide switch for on-off. Any voltage from 1.5V to 9V will power the injector. You don't have to bother installing a battery holder in the unit since the power consumption is so low it will last for years without a battery replacement. Simply solder the power leads to the positive and negative battery terminals. Use a high wattage gun (see Section 4-3) to make the connections quickly so the battery is not exposed to excessive heat. Once the injector is completed you can check its operation with an audio amplifier or oscilloscope. An approximate 1.5 KHZ tone will be heard from the amplifier's speaker or a waveform similar to Figure 8-1C will be seen on your oscilloscope.

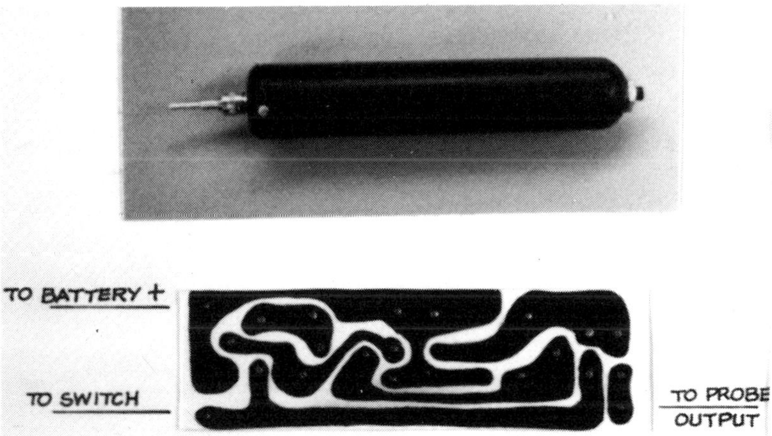

Figure 8-1B. Signal injector probe and exact scale PC layout.

Figure 8-1C. Signal injector waveform.

Now we are ready to begin troubleshooting with the signal injector. Refer to Figure 8-1D while reading the following discussion. The RF and IF portions of the radio are contained in a single IC; the AF portion in separate transistor stages.

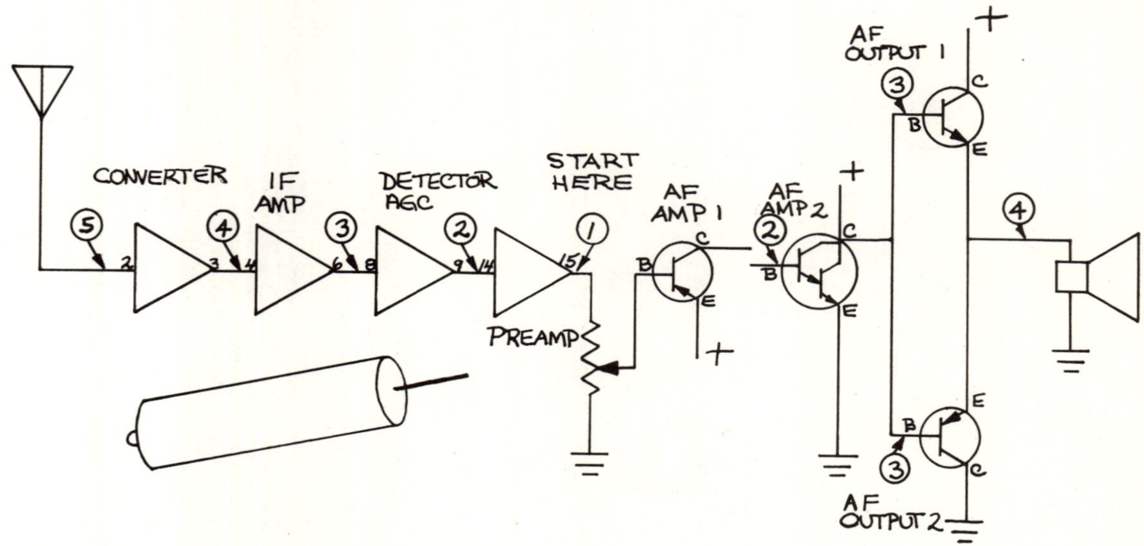

Figure 8-1D. Troubleshooting an AM radio using a signal injector.

Suppose the radio is dead. You have already checked the battery or power supply and it's okay. Here's the ideal place for . . . you guessed it . . . a signal injector. When injector troubleshooting in a radio, it's the best policy to start checking from the midpoint of the signal path. With one injection the radio can be divided in two sections localizing the trouble to one half or the other. The top of the volume control makes a perfect place to start; it's easy to locate and divides the RF-IF section from the AF stages. This test gives the following information:

A. No injector tone means the audio section is defective. Further troubleshooting in this area is required.

B. Good tone, regulated by the volume control means the AF section is okay. Further checking in the RF-IF section is the next step.

To further check the audio section simply touch each transistor base with the injector probe as you move from the volume control towards the speaker. When the tone is heard for the first time, the stage checked previously is the defective one. For example if the tone is first heard when injecting at the base of AF output 1, then the trouble will be in the AF amplifier 2 stage.

On the other hand, when the AF section is good and you are signal injecting the RF-IF stages, touch the probe to each IC input as you move towards the antenna. When the tone is no longer heard, that stage is defective. Suppose you hear the tone at IC pin 14 and IC pin 8 but lose it at IC pin 4. The defective stage is the IF stage.

Once you have the bad stage localized (using the signal injector it takes less than a minute in a familiar receiver) then further checking in that stage with a meter for proper voltages and resistances must be made to nail down the defective part.

A final point to remember when signal injecting is that it's normal for the speaker volume to decrease when checking the IF and RF stages as compared to the audio stages. Sometimes this creates a confusing situation. Since more amplifying stages are involved it logically seems the tone should become louder. The reason for lower volume is that only the harmonics of the signal injector square wave are being amplified, and they are much lower in amplitude than the fundamental audio frequency.

8-2. SWEET AND SIMPLE TRANSISTOR-DIODE CHECKER

This transistor-diode checker is a honey. You don't have to know a great deal about the transistor to be checked. It's helpful if you know whether it's PNP or NPN and the lead configuration, but it's not absolutely essential. You can't hurt the transistor or the checker if you have it connected wrong. Just try all possible lead arrangements until the transistor checks okay. If it doesn't check out it must be bad.

A good low power transistor will light one LED if the transistor is good; the more gain the brighter the LED. If the transistor is leaky the LED will glow slightly. A good power transistor will light three LED's. The brightness of the LED's is an indication of the relative gain. Checking a diode or rectifier will light an LED in the forward direction and keep it off in the reverse direction. These are the operation instructions in a nutshell . . . nice and simple.

The circuit for the transistor-diode checker is shown in Figure 8-2A. A step-down transformer feeds a half-wave rectifier and filter circuit for the power supply. The transistor to be tested is placed into a common emitter amplifier circuit. Leakage of the emitter-collector junction will be indicated by the LED brightness when the leakage-gain switch is in the leakage position. Depressing this switch will apply forward base-emitter bias to the transistor. A good transistor will turn on causing current flow in the emitter-collector circuit indicated by the brightness of the LED. One DPDT switch is used to change power supply polarity for PNP and NPN transistors and the other switch increases the LED load for power transistors.

Figure 8-2B shows the finished transistor-diode checker. Try to mount the LED's in a recessed area of the chassis so their brightness levels will not be obliterated by room light. After the construction is completed check a bunch of good transistors and diodes to become familiar with the checker's operation and idiosyncrasies.

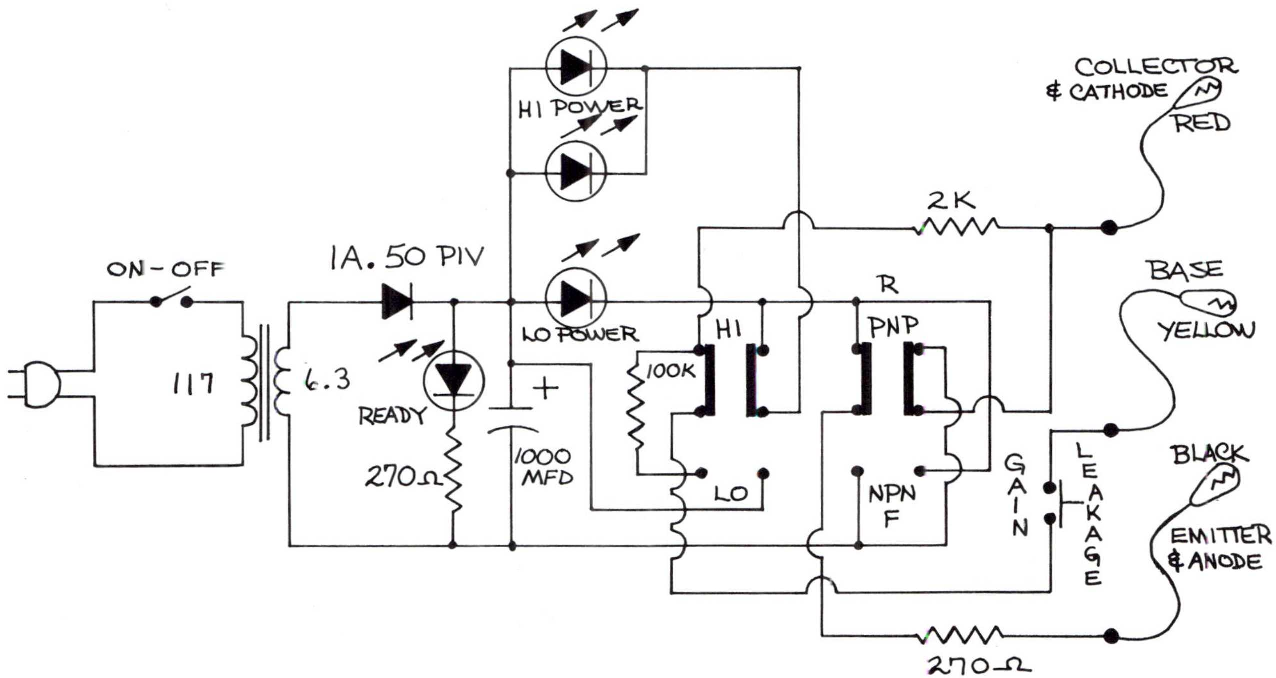

Figure 8-2A. Transistor-diode checker schematic.

Figure 8-2B. Transistor-diode checker.

8-3. A TESTER FOR SCR'S AND TRIACS

The SCR and triac tester schematic shown in Figure 8-3A is simple and inexpensive to build and really a timesaver to use. The entire project can be easily completed in one evening.

The checker can be installed into a small minibox. The wiring layout is not critical. Mounting the #47 lamp in a vinyl grommet will save a socket assembly and still make a nice appearance. Color code the tester leads for easy identification. Be certain to use a vinyl insulator over each test lead clip to prevent accidental short circuits.

Here's how to check an SCR. Connect the anode, cathode, and gate leads to the appropriate tester leads. Place the polarity switch in the negative position. The lamp should be off. Press the gate switch briefly. The lamp should still be off. Flip the polarity switch to the positive position—the lamp should remain off. Now press the gate switch for a moment. The lamp should light and remain on even after the gate switch is released. Any SCR that deviates from this prodecure is probably bad.

Checking a triac is similar to checking an SCR. The only difference is that the triac will gate on with a negative as well as a positive polarity. After the triac is connected to the appropriate tester leads, flip the polarity switch to the negative position. The lamp should be off. Press the gate switch briefly. The lamp should turn on and remain on when the gate switch is released. Change the polarity switch to the positive position—the lamp should go off. Briefly push the gate switch again.

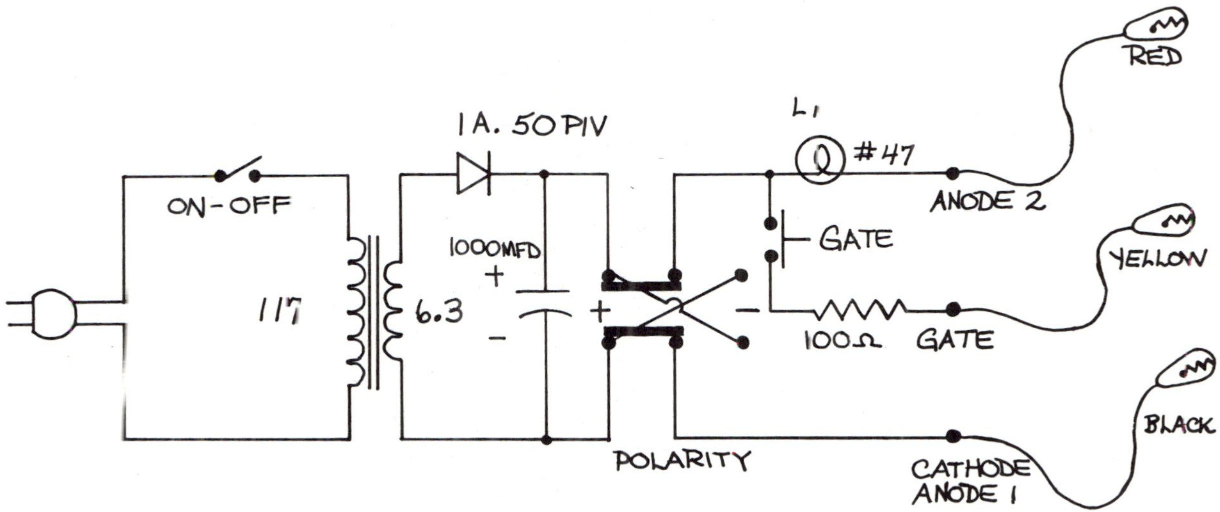

Figure 8-3A. SCR and triac tester schematic.

The lamp should light and remain on. This completes the triac test.

Figure 8-3B shows the completed SCR and triac tester. You might want to attach the operation tables to the minibox for future reference. The tester anode and cathode leads can also be used to check continuity in low resistance circuits. Simply place the tester leads across the wire or part to be checked. A low resistance will light the bulb normally. As the resistance increases the lamp will become proportionally dimmer.

SCR CHECK

SWITCH POS.	GATE OFF	GATE ON	GATE OFF
+	L1 OFF	L1 ON	L1 ON
−	L1 OFF	L1 OFF	L1 OFF

TRIAC CHECK

SWITCH POS.	GATE OFF	GATE ON	GATE OFF
+	L1 OFF	L1 ON	L1 ON
−	L1 OFF	L1 ON	L1 ON

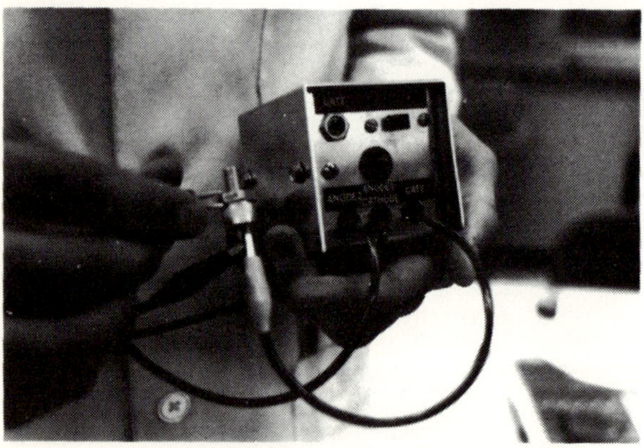

Figure 8-3B. SCR and triac tester.

8-4. A FOOLPROOF DEVICE FOR CHECKING UNIJUNCTION TRANSISTORS

You'll like this unijunction transistor tester because it will sound off when the transistor is good. A defective unijunction will be as quiet as a mouse.

Figure 8-4A shows the schematic of the unijunction transistor checker. A voltage doubler circuit supplies power to an oscillator pulse circuit. A good transistor will cause the circuit to become operational emitting sound from the 2″ speaker.

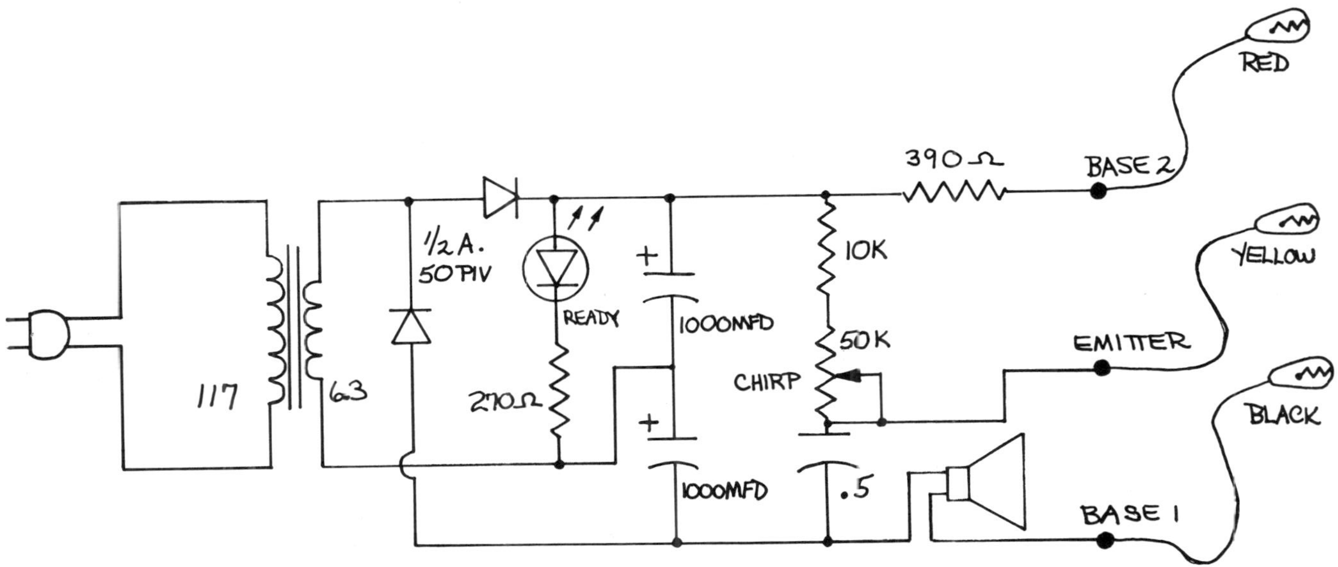

Figure 8-4A. Unijunction transistor checker schematic.

To check a unijunction simply connect its leads to the appropriate tester clips. Turn the tone control until audio is heard. If you don't know the lead identification just try all the combinations until the speaker "talks." You'll probably hear audio for either of the base lead combinations. When B_1 test lead is connected to the UJT B_1, the chirp will be the lowest frequency. Nothing will be damaged by connecting the right leads to the wrong clips. A silent unijunction is a bad unijunction . . . easy as pie.

Refer to Figure 8-4B for a peek at a completed unijunction transistor checker. If you look closely you can almost hear it talking.

Figure 8-4B. Unijunction transistor checker.

8-5. BUILD A VARIABLE LOW VOLTAGE, HIGH CURRENT, POWER SUPPLY

This variable regulated DC power source will supply current up to 2 amps at voltages from .5 to 22V. There's plenty of power to operate a car radio or tape player on the bench. For a circuit with so few parts the regulation is surprisingly good as shown in Figure 8-5A.

The heart of the power supply is the 2N174 170W power transistor. The transistor and heat sink assembly are shown in Figure 8-5B. This assembly can often be found in surplus stores or mail order catalogs for less than two dollars. The transistor is connected in the circuit as a linear series voltage regulator, see Figure 8-5C.

The transformer steps down the AC voltage to 28V @ 2 amps. A 2A bridge rectifier changes the AC to DC. The 2,000 MFD capacitor acts as the input filter capacitor. The 2 amp fuse must be placed after the input filter capacitor since it would be

Figure 8-5A. Regulation characteristics of the DC power supply.

Figure 8-5B. Transistor and heat sink assembly.

blown by the initial capacitor surge. The transistor acts like a variable resistance in series with the load. The 1K potentiometer controls the transistor conduction from cutoff to saturation depending on its setting. When the pot wiper is at the top of the schematic the transistor is turned off. The zener diode limits the output voltage to 22 volts and provides the reference voltage for the base element of the transistor. Output filtering is provided by the 1,000 MFD capacitor.

Construction of the supply requires no special wiring techniques. Be sure that the transistor or the heat sink is electrically insulated from the chassis. Use fiber shoulder washers and a mica insulating wafer to insulate the transistor from the heat sink. Be sure to use silicone grease between the transistor, mica wafer, and the heat sink to insure maximum heat transfer. When the supply is operated at low voltages and high currents, the transistor and heat sink must dissipate a large amount of power. Use 18 gauge wire for the part of the circuit shown in heavy black lines in Figure 8-5C.

When the power supply project is finished you'll be pleased with the results. Figure 8-5D shows the finished supply built into the cabinet described in Section 10-3. This particular project has the AC power supply described in the following section (8-6) included in the same cabinet.

8-6. PLANS FOR A TRIAC CONTROLLED VARIABLE AC SUPPLY

The variable AC power supply shown in Figure 8-6 will supply AC voltage from zero to 117 VAC @ 8A. Use this supply for a lamp dimmer, portable electric tool speed controller, or whenever a variable AC voltage is needed.

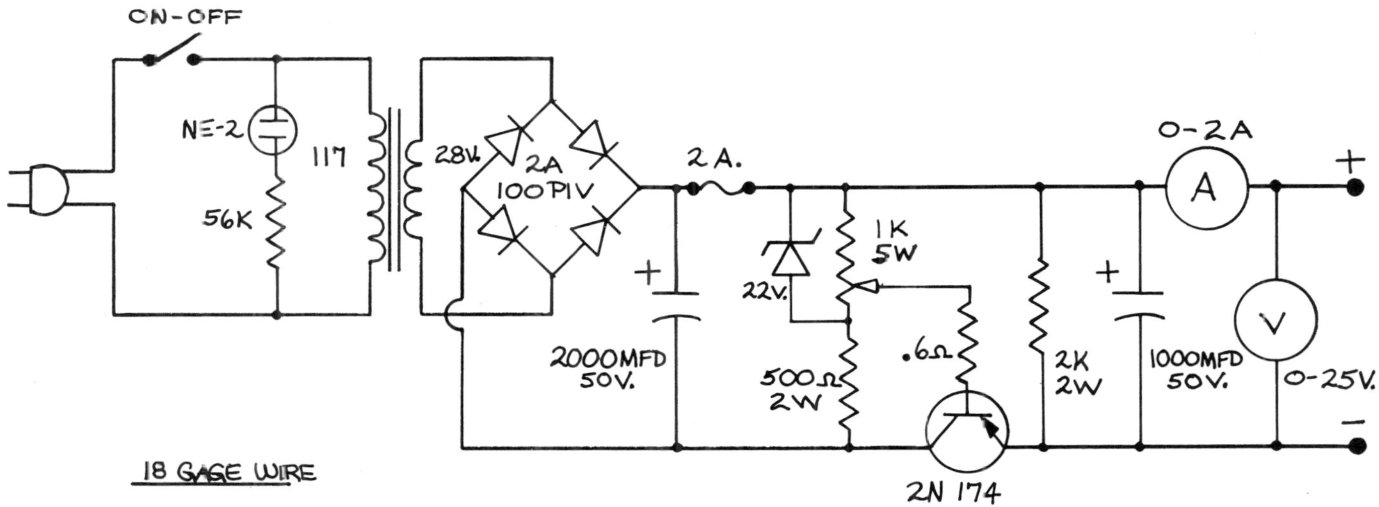

Figure 8-5C. DC power supply schematic.

Figure 8-5D. Variable AC-DC power supply.

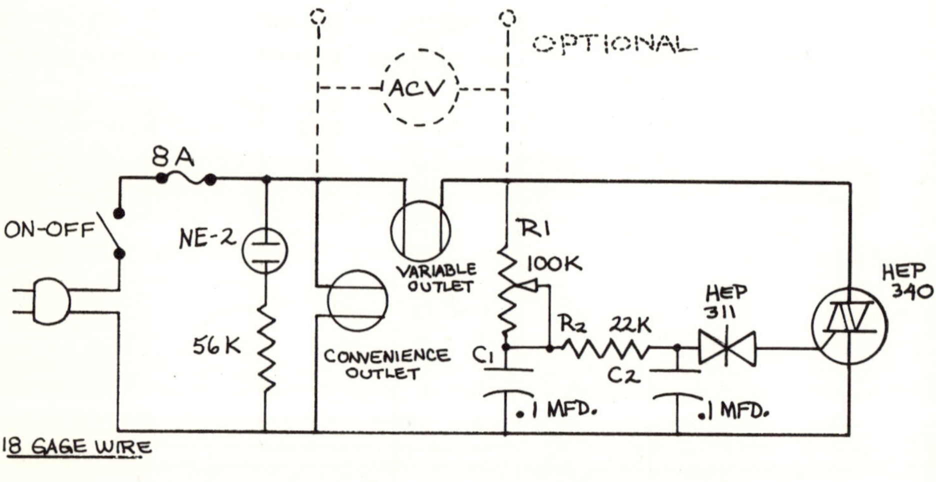

Figure 8-6. Variable AC power supply schematic.

The triac is a full-wave gate-controlled thyristor. It has the ability to conduct on the positive or negative pulse of each AC cycle. The gate can trigger the triac on for a small part of each cycle, the complete cycle, or any part in between depending on the potentiometer setting. R_1 and C_1 form a phase shifting network to supply varying phase shifts to the gate circuit. Diac 311 is a bidirectional trigger device for the triac. C_1, R_2, and C_2 provide a double time constant filter circuit for smooth uniform turn-on and turn-off conditions.

Heat sink the triac as described in Section 8-5. Many triacs are electrically insulated from their metal case so mica insulating wafers and fiber shoulder washers are not needed for mounting. However if the triac is not insulated, one side of the line potential will be present on the metal case. An ohmmeter check will confirm continuity between any triac element and its case. Be absolutely certain that there's no continuity between the triac's elements and the chassis after the triac is mounted.

Use 18 gauge wire or heavier to hook up the load current part of the AC power supply (shown in darker lines in Figure 8-6). You also may want to include banana jacks in parallel with the variable AC socket to provide easy access for test leads. Caution: be careful never to short circuit the variable socket, or full line current will zap through the anodes of the triac. Generally the triac will self destruct before the fuse. Another nice feature to include in this project is a convenience outlet for standard 117 VAC line voltage operation. An AC meter to monitor the variable AC outlet is a worthwhile added feature. However, if you do not include a meter, make certain to calibrate the potentiometer, R_1, with front panel markings to indicate approximate voltage settings.

9 How to Clean and Recondition Electronic Equipment

Restoring electronic equipment to like-new condition inside and out is discussed in this section. You'll find several methods described that effectively remove dirt and grime from electronic circuitry. Preventive maintenance techniques are covered—they will extend the life of any electronic instrument.

If you ever wanted to restore a front panel, metal case, or housing, Section 9-3 offers tips and tactics that will guarantee success. Before testing and recalibrating your equipment be sure to read Section 9-4. You may save yourself a lot of time and work.

9-1. THERE'S ELECTRONIC CIRCUITRY UNDER ALL THAT DIRT

If you ever looked in the back of a color TV that has been operating for a number of years, you know the amount of dirt, dust, film, and residue that can accumulate on the circuitry. TV's are not the exception. All electronic equipment is exposed to airborne dirt and contamination; the dirtier the environment, the dirtier the electronics.

Dirt contaminations may contain carbon, sulphur, metal oxides, lead, and various other particles. Most contaminants will absorb moisture, forming high resistance leakage paths whose conductivity is a function of atmospheric conditions. Just imagine taking a handful of meg ohm resistors and indiscriminately soldering them from point to point throughout an electronic circuit. No self respecting technician would ever think of doing this, but dirt and moisture are accomplishing the same result.

Needless to say thorough cleaning of electronic apparatus can often do wonders in correcting a multitude of problems caused by high resistance paths. Not only will apparent defects be cured but often time consuming calibrations and adjustments will be unnecessary.

What's the best method to clean electronic equipment? Effective cleaning is always accomplished by a mechanical and chemical combination. Here's a list of common cleaning methods:

1. Brush and solvent technique
2. Ultrasonic baths
3. Pressure-spray methods

The first method is the most common. Usually the chassis is vacuumed and swept with a soft brush. After the loose dirt is removed the brush, cotton swab or whatever is dipped in alcohol or other suitable solvent. Each part is then literally scrubbed and dried. This is a very time consuming process and never 100% successful.

The second method, ultrasonic baths, uses a solvent agitated by ultrasonic waves. In most cases this method is only practical for small devices. Then there's the chance that the ultrasonic frequencies could damage an electronic component.

The pressure-spray methods of cleaning are the most practical. They do an excellent job in minimum time. Figure 9-1A shows a combination aerosol spray can equipped with an extension spray brush. This particular brush called the Cobra—is really neat. The brush is connected to a three foot nylon tube which snaps on the valve of the aerosol can. Exerting side motion on the brush causes the cleaning solvent to spray out of the brush. Once the parts have been sprayed much of the dirt will wash away. Stubborn grime can be brushed loose. Need more solvent? Just exert side pressure on the spray brush for new solvent while you're scrubbing. It works like magic. When the cleaning is finished pull off the extension spray brush and reinsert the regular spray valve on the aerosol can. Shoot solvent over the parts again to flush the last remaining residue away.

Courtesy of Miller-Stephenson Chemical Co.

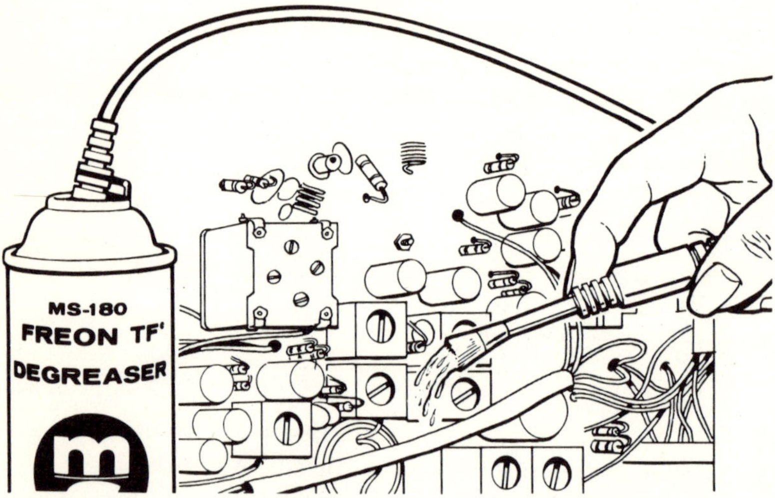

Figure 9-1A. Aerosol degreaser and extension spray brush.

Did you ever think about cleaning electronic circuits with soap and water? Sounds wierd. Yet one of the most effective cleaning systems available uses water, air pressure, and a specially formulated cleaning solution. Figure 9-1B shows a professional cleaning system. The equipment to be cleaned is placed on a rotary turntable inside the cleaning booth. Water and cleaning solution are fine-sprayed under pressure on the electronic apparatus. After the cleaning action is completed, the circuits are pressure rinsed with water until all detergent bubbles have disappeared. Excess moisture is blown away with air pressure. It's as simple as that. The only catch to cleaning electronic equipment with this method is that thorough drying is essential. A drying oven shown in Figure 9-1B does the job. This low temperature, high air velocity oven will remove every speck of water with no possibility of overheating.

Every electronic shop should have a cleaning system like

Courtesy of M. P. Odell Co.

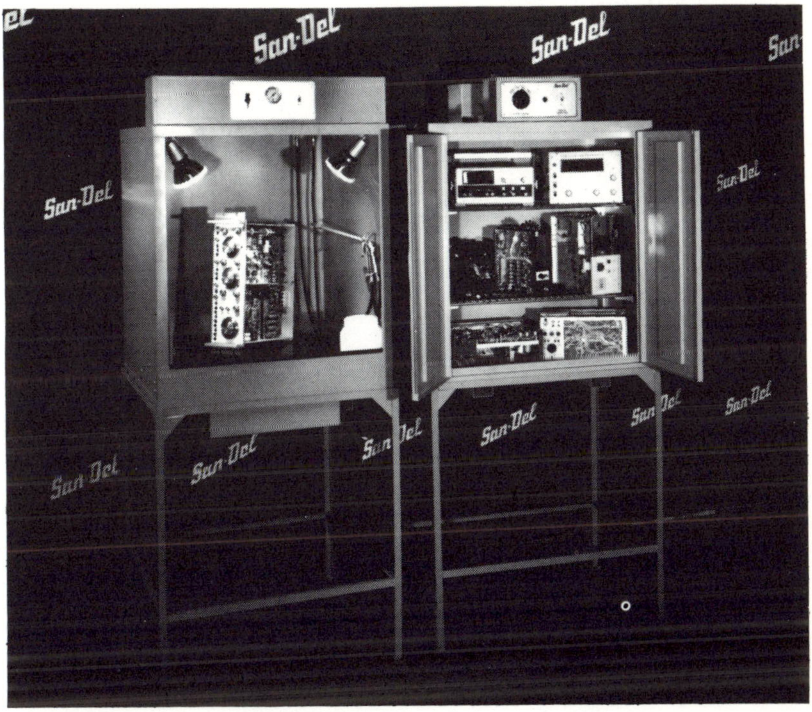

Figure 9-1B. Professional electronic apparatus cleaning system.

the one just described. Most don't—they are expensive. However, electronic apparatus can be cleaned effectively in a similar manner on a small scale with minimum equipment. Here's what is needed:

1. Vacuum cleaner
2. Water
3. Liquid detergent for washing dishes
4. Pressure sprayer
5. Heat lamp or hair dryer

Vacuum all the loose dust, dirt, and debris from the unit. Use a soft paint brush to loosen up the dirt. Mix up a cleaning solution containing one part of dish washing detergent to nine parts of water. Set the electronic apparatus in a laundry tub. Make certain that the detergent solution can not come in contact with anything that could be damaged by water such as paper labels, cardboard covered capacitors, speaker cones, meters, wooden cabinetry, etc.

Various devices can be used for spraying depending on what is available. An electric air compressor-sprayer would be ideal. A pressurized garden type sprayer also works well. Even a pump action oil can will work although not as effective as the sprayers.

Spray the detergent water cleaning solution throughout the electronic apparatus until it shines like new. Rinse spray with clean water until all detergent bubbles disappear. Air spray to remove excess water and moisture. Immediately dry with a heat lamp(s) or blow gun hair dryer. If it is a hot sunny day the apparatus can be air dried outside in the sunshine. Be certain to dry the apparatus at once so moisture damage is prevented and no rust can form. When drying with a heat lamp or hair dryer be sure to keep moving the heat source so no damaging hot spots develop on the equipment.

Figure 9-1C shows a liquid cleaning concentrate solution for use with tap water that is made especially for cleaning electronic equipment. It is an economical concentrate safe for both electronic equipment and for the operator. It will remove dirt, dust, oily residue as well as deoxidize and brighten switch and

Courtesy of M. P. Odell Co.

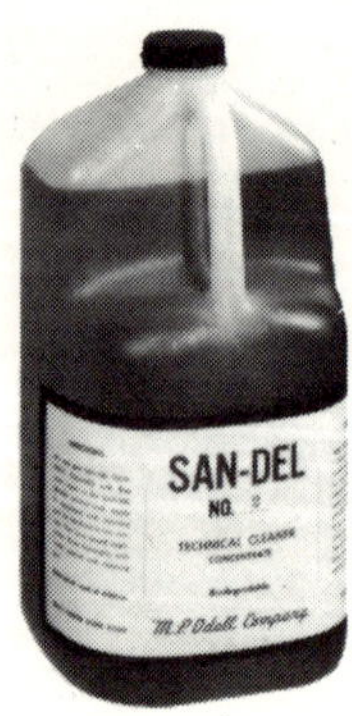

Figure 9-1C. Liquid cleaning concentrate.

wiper contacts. This cleaner is available from M. P. Odell Co. 26612 Center Ridge Rd. Cleveland, OH 44145.

9-2. PREVENTIVE MAINTENANCE THAT PAYS BIG DIVIDENDS

Give your electronic equipment a little attention before a catastrophic failure occurs. Most electronic apparatus deteriorates so gradually that substandard performance is often overlooked. You will be pleasantly surprised to see how preventive maintenance can make your instruments perform and look like new again.

Let's illustrate some of the things that can be done using a meter type instrument such as a VTVM. Take a good look at your old standby. It's probably dusty and dirty. Maybe a knob is loose or missing. The meter face is probably smudged, scratched, or dirty. Do the range and function switches rotate smoothly and positively click into position? Inspect the test probe and common lead as well as the line cord. You'll probably find deterioration that has just been ignored. The VTVM sits kind of cock-eyed probably because of that missing foot.

Set some time aside to recondition the VTVM. It will probably take a lot less time than you think. Start with a good cleaning of the case using liquid household cleanser, Figure 9-2A.

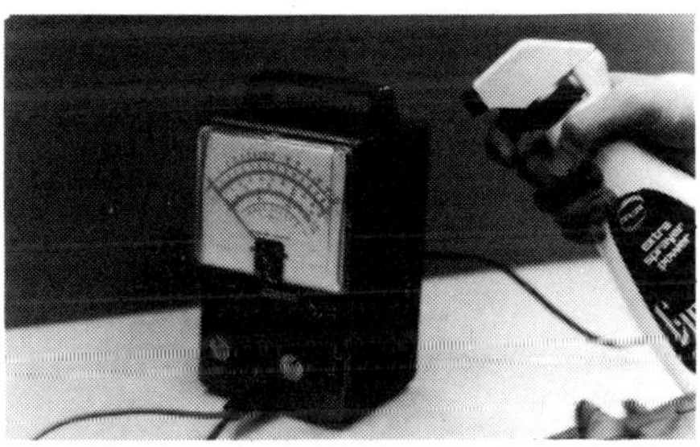

Figure 9-2A. Spray cleaning a VTVM case.

The type in a plastic spray bottle works like magic. Remove the knobs and plug-in leads before spraying. Let the cleaner loosen the dirt before wiping with a clean soft cotton cloth. A section of an old sheet does the job. Finish by waxing and polishing. Next spray the knobs and scrub them with a discarded tooth brush. Rinse with hot water and they will sparkle like new. While the knobs are off check all the nuts that keep the switches and pots in place for tightness. They have a tendency to become loose after constant use. A nutdriver or box wrench is the tool for this operation.

Spend some time with the meter face. Most plastic faces will snap off the instrument with a little careful prying. Wash both sides of the face with window washing liquid. Minor scratches can be removed with plastic lens cleaner available at any eye-glass store. Snap the face back on the meter carefully lining up the zero adjustment screw with the meter mechanism. Readjust for mechanical zero while the VTVM is in its normal physical position.

Remove the case to relubricate the detent mechanism on the rotary switches with white grease. A small dab should do the trick. Next spray all switch contacts with tuner or contact cleaner. Rotate the switches back and forth a number of times to help clean and burnish the contacts before the cleaner evaporates. Shoot some contact cleaner into the zero and ohms adjust pots. Turn the pot shafts back and forth to help the cleaning action.

Inspect all the wires that are fastened to the switches and pots. Actually touch them with a pencil to make sure they are fastened securely. Often these wires can become brittle and break off due to the constant vibration of the switching motion. Make a careful visual inspection of all parts looking for discolored or charred components, especially the precision resistors mounted on the rotary switches. If one or more of these resistors is open or out of tolerance they will have to be replaced. Don't try to unsolder them from the lugs on the wafer switches or the lugs may break. Simply cut the resistor out leaving its leads intact. Fasten the new resistor on to the old resistor leads, see Figure 9-2B.

While you are inside of the VTVM change the battery. Recalibrate the instrument as per instructions in the service manual. If the line cord or test leads have had it, replace them, see Section 7-4; they can be flexed just so many times. Solder a new insulated alligator clip on the common lead. The chances are the old one is worn out. Eyeball the rest of the VTVM looking for anything else that needs repair or replacing. Pop the case back on and fire up the instrument. It looks and performs like new!

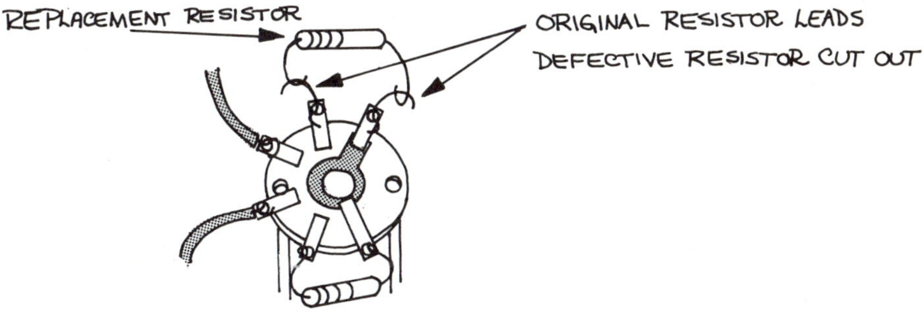

Figure 9-2B. Mounting a replacement resistor on a wafer switch.

You'll probably be so happy with the results that all your instruments will get the same kind of treatment. Sure it takes a little time but comparing the cost with new instruments makes it all worthwhile. Besides, it makes a person feel better working with tip-top equipment.

9-3. CABINET RECONDITIONING TECHNIQUES

Electronic enclosures get nicked, scratched, and bumped like any well used apparatus. There are various methods that can be used to get these cabinets back in first class condition.

Most electronic cases are made from some type of sheet metal or plastic. Many times a good cleaning, waxing, and polishing as described in Section 9-2 will be all that's necessary. Slight nicks and scratches can be satisfactorily fixed with an artist's brush and touch-up paint.

When cleaning won't do the job or maybe you want to match all your instruments with the same color scheme, then repainting may be the answer. Remove the knobs and all easily removable items. Clean, fine sand, and dust the cabinet. Small dents can be filled in with spot putty available at automobile parts stores. Before repainting be sure to make a record of all labels that will have to be redone. Mask all pilot lamps, switches, fuse holders, dials, meters, jacks, etc. that can't be easily removed. Aerosol spray paint will really do a super job to restore the finish. Start with a primer paint for good adhesion. If you're painting over aluminum be sure to use a zink-chromate primer.

It's best to practice spraying on a scrap before actually painting the instrument case. Aerosol paints are different in consistency so it pays to make your mistakes on a practice piece. Always spray the least prominent section first; save the top and front panel for last. This will prevent any overspray or smudges from occurring on the surfaces that are really obvious. Spray in such a way that the nozzle always remains the same distance from the cabinet. If you are too close the paint will run; too far and the paint will have a sandy texture. Try to complete the entire spray operation before the paint begins to dry. Remove the masking tape before the finish coat has set.

After the paint has had a chance to dry thoroughly relabel the enclosure. The most professional looking labels can usually be made with the pressure transfer type, Figure 9-3. This labeling method is easy to use and works very well. Follow these steps for a perfect labeling job:

1. Position the label to be transferred on the cabinet. Keep the backing sheet under the transfer sheet but slide it down below the label.
2. Press lightly on the labels with a dull pencil point or other dull pointer. Press just hard enough to transfer the label. The label will hairline crack if excessive pressure is used.
3. When the label has been transferred lay the backing sheet over the labels and burnish lightly to permanently set them in place.
4. After all labels are completed spray them with clear acrylic spray.

Reconditioning wood enclosures is another story. First try cleaning the wood with a combination cleaner-waxer-polish. Often this is all that's necessary to bring back the original wood

Figure 9-3. Labeling an electronic enclosure with pressure-sensitive labels.

luster. If this won't do the job, the best alternative is to remove the old finish and redo. Use a water rinsable paint stripper. Liberally brush the paint remover over the wooden surface. Allow enough time for the old finish to become soft and removable before scraping it off. Two or more stripper applications may be necessary before all the original finish is removed. Water rinse or hose to remove the last of the remover and finish. When working with plywoods or thin veneers keep the water to an absolute minimum and dry with a towel or sponge as quickly as possible.

After the wood has completely dried, sand the surface as needed. Sand with the grain, finishing up the sanding with fine paper. When the sanding is completed be sure to remove every speck of dust with a tack rag or lintless rag dampened with turpentine. If you are going to stain the wood to enrich the color or reinforce the grain pattern be sure to use ready mixed oil stains. Apply with a cloth after you have practiced on a hidden area. Wipe off excess stain as soon as the color penetration is achieved.

Finish the wood with a combination sealer-finish available in most paint stores. Three coats are usually necessary to restore that factory shine. These all-in-one finishes dry to a medium luster looking much like fine hand-rubbed furniture. Just follow the directions on the can and you will be pleased with the results.

9-4. METHODS FOR TESTING AND RECALIBRATION

All electronic equipment deteriorates with age. Much of the deterioration is the result of airborne dirt and grime. So before going through recalibration procedures outlined in the service manual it is a wise policy to first clean the parts and circuitry especially if there's any evidence of dirt, refer to Section 9-1.

Next clean and recondition all switches, pots, and jacks as described in Section 9-2. At the same time keep an eye out for any visible evidence of defective parts such as overheated resistors, broken capacitors, broken leads, etc. Once the equipment is clean and all mechanical contact surfaces reconditioned, try operating the instrument. Run it through its specifications. You'll probably find out that the instrument is working fine again and no further work is necessary.

If the instrument is slightly off its specs, now's the time to recalibrate and tweek up the adjustments as described in the manufacturer's data. Don't try to recalibrate by hit or miss methods. Don't use one manufacturer's directions to adjust some other company's instrument just because they are similar, refer to Section 3-4. Most electronic instruments are unique in one way or another. Save yourself a lot of time and trouble by always using the correct service information.

One of the big hang-ups when calibrating test equipment is trying to find an accurate standard reference voltage, current, or frequency source. Most technicians don't have laboratory standards sitting around their workshop. A new carbon-zink battery (flashlight cell) is an accurate source for DC voltage; it produces 1.55 VDC. Using a flashlight battery in series with a 1% precision resistor will make a satisfactory circuit for checking DC current, Figure 9-4A. Just be sure to draw minimum current from the battery so it won't be loaded down. Here's an example:

$$
\begin{aligned}
{}^{E}\text{Batt.} &= 1.55\text{V} \\
\text{Resistor} &= 1\text{K}\Omega \\
\text{Current} &= \frac{1.55}{1{,}000} \\
\text{Current} &= 1.55\ \text{MA}
\end{aligned}
$$

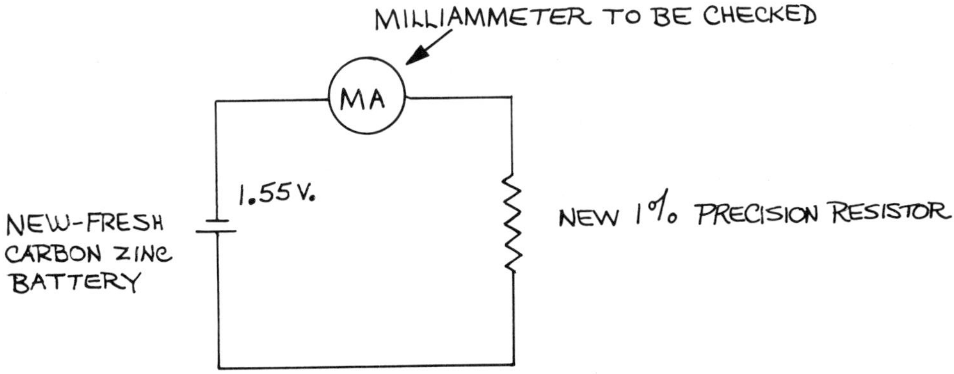

Figure 9-4A. Circuit for checking DC current calibration.

Ac calibration voltage is tougher to find. You can't use line voltage since it normally fluctuates. If you have a recently calibrated DVM or can obtain access to one, your AC problem is solved. Many DVM's are accurate within ±1% for AC voltage and current. This is satisfactory accuracy for calibration of the majority of instruments. Simply monitor line voltage with the DVM to obtain an accurate AC measurement. Now line voltage can be used for AC calibration. If lower AC voltage is needed just connect a resistive voltage divider network across the line to pick off any voltage you desire, see Figure 9-4B. Suppose the DVM measures line voltage of 120 VAC. The voltage drop across points A & B, B & C, and C & D, will be 1/3 of line voltage or 40 VAC. The voltage between A & C will be 80 VAC. For this particular circuit an AC milliammeter should read 8 MA. One-half watt resistors will be adequate for this temporary calibration circuit. Follow this same procedure to obtain any AC voltage or current you desire. Try to keep the total resistance value as low as possible with reasonable wattage values. This will tend to prevent erroneous measurements when calibrating low ohms per volt equipment.

For accurate low frequency calibration, line frequency can be used as the reference. Unlike line voltage, line frequency is kept on the button. For instance, suppose you're checking the calibration of an audio generator. First obtain two cycles of the line AC waveform on your oscilloscope. Set the scope controls so that each cycle uses a fixed amount of distance on the scope graticule; one inch or ten centimeters per cycle would be ideal. Without moving any of the scope frequency or positioning controls, substitute the audio generator for line voltage. Turn the generator frequency knobs until two cycles are again shown on the scope. If the generator is calibrated correctly the dial should be reading 60 HZ.

Any radio or TV station's carrier frequency can be used as

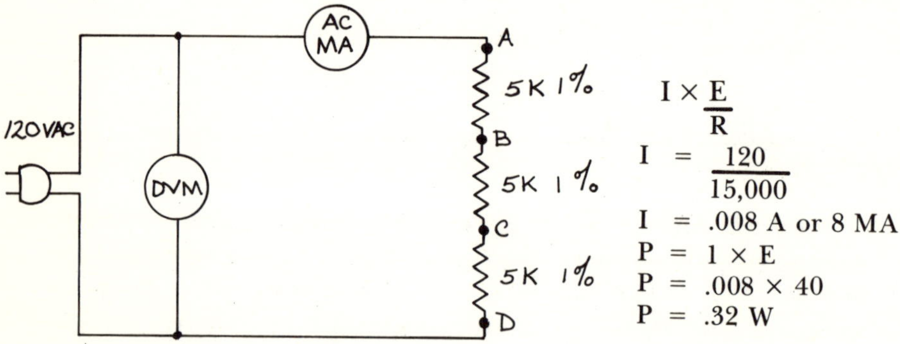

Figure 9-4B. Resistance voltage divider network for an AC calibration reference source.

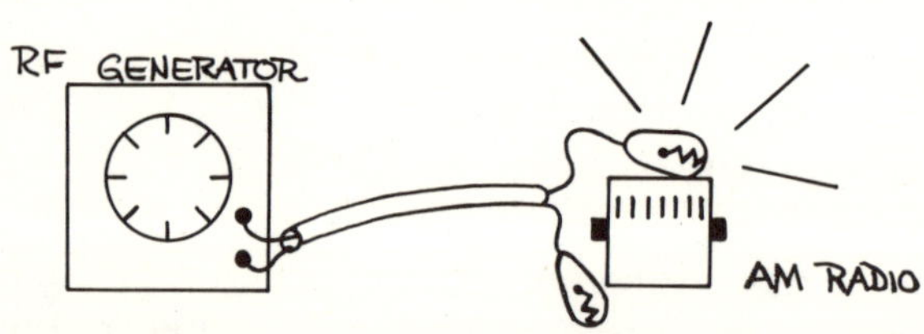

Figure 9-4C. Zero beating an RF generator with an AM radio for frequency calibration.

a frequency standard. By law their broadcasting frequency must be held within very rigid limits. You can easily check a generator RF output by zero beating it with a radio station. Tune a radio to a common station in your area, say 1100 KHZ. Now adjust the RF generator to the same frequency while keeping the generator's output cables physically close to the radio antenna. When the two signals match you'll hear a zero beat in the radio speaker. If the generator is good its dial should read 1100 KHZ, see Figure 9-4C. For further frequency calibration techniques read about the digital frequency counter in Section 12-2.

10 Fast, Easy Ways to Build Electronic Enclosures

If you have ever bought a factory made electronic enclosure or have priced them in an electronic catalog, you know that they are quite expensive. This section deals with this problem by showing how to work successfully with sheet metal. All the designs revealed on the following pages can be constructed with just a few sheet metal tools.

Start out by building a cowl-type minibox. There are a million uses for these little boxes. Plans for an easily constructed all-purpose chassis are outlined in Section 10-3. Here's a sharp looking enclosure that fits the bill for a multitude of electronic projects. If you're interested in a classy console cabinet for that special instrument you're planning to make, the blueprint offered in Section 10-4 may be just what the doctor ordered. Want a really different type electronic chassis? Be sure to check the unusual design shown in Section 10-5.

10-1. WORKING TIPS FOR SHEET METAL ENCLOSURES

Part of the fun of building an electronic construction project is designing and fabricating the chassis enclosure. A well thought out design skillfully constructed brings a good deal of pride and satisfaction to the builder.

Sheet metal is an ideal chassis material for a number of reasons:

1. Relatively inexpensive
2. Easy to mount controls and parts
3. Shielding effect
4. Relatively easy to fabricate
5. Ease of finishing
6. Minimum bulk
7. Maximum strength

Choose 22 gauge cold rolled steel or 16-20 gauge semihard aluminum. The aluminum will be more costly but somewhat easier to work with. After you have made several sketches lay out the design on cardboard. The cardboard should be approximately equal in thickness to the metal. Cut and bend the cardboard with the same tools that you will use with metal. Remember if you have trouble bending or cutting the cardboard, it's going to be super difficult with sheet metal. A good chassis design should have all outside and straight cuts, and be easy to bend with the tools available in your workshop. A small box and pan brake and a bench mounted sheet metal squaring shear makes fabrication a breeze, see Figure 10-1A. If you don't have these tools don't dismay; common hand tools can also do the job.

Try to stay away from designs that require almost perfect sheet metal work. Functional designs can cover up a multitude of sheet metal mistakes and so can covers and flaps. Keep the easiest-to-make section of the chassis on the outside and the rest of the chassis will look better. Allow enough interior space for working area during construction. Mount all the electrical parts on one section of the chassis. Use the other section for a dust cover. You'll be glad not to have interconnecting cables between enclosure sections.

After the cardboard model is completed, scribe layout lines on the metal with a scratch awl. Make certain these lines will end up on the inside of the enclosure where they won't show. Refer to Figure 10-1B, a partial view of a typical chassis layout, for the following discussion. Include flaps on all edges of the chassis. They will provide structural strength as well as making

Figure 10-1A. Sheet metal fabricating tools.

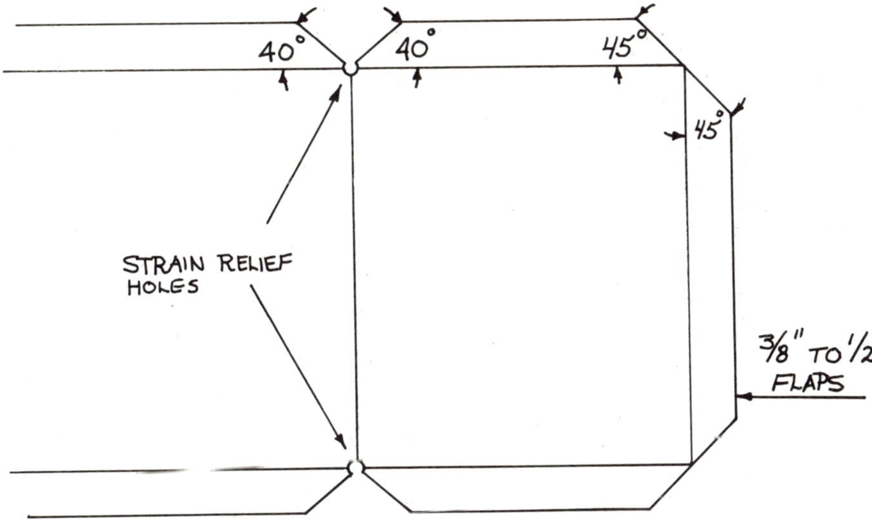

Figure 10-1B. Typical chassis pattern (partial view).

a nice appearance. Drill small holes, 1/8″, at all corners to provide strain relief during bending. It's a good idea to lay out the corner angles less than 45°. The reason being that when the metal is bent 90° the flaps will have no chance of touching each other causing a buckle; 45° angles work only for perfect craftsmen. Be sure that all cuts, holes, and openings are made before the metal is bent. Refer to the author's Electronic Technician's Handbook of Time-Savers And Shortcuts for more detailed information on chassis construction techniques.

Begin the bending operation by bending all of the flaps first. When the flaps are finished it is generally best to bend angles less than 90° first and finish the bending operation with the 90° bends. When a box and pan brake is not available, satisfactory bending can be accomplished with hardwood strips and C clamps. Cut the 1″ × 1″ wood strips the exact length of the bend. Clamp the strips securely above and below the bend as shown in Figure 10-1C. Using another strip of wood, bend the metal to the desired angle. Follow this procedure for the remainder of the bends.

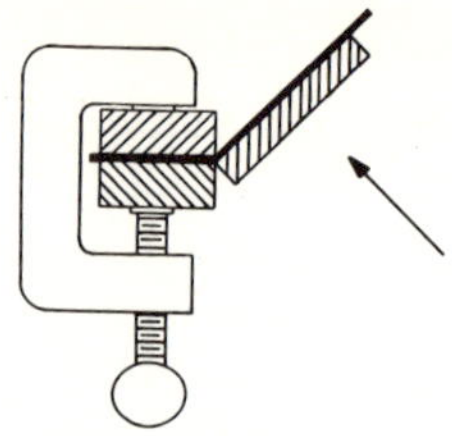

Figure 10-1C. Bending sheet metal using wood strips.

10-2. SPEEDY PLANS FOR A COWL-TYPE MINIBOX

The cowl-type minibox is fast and easy to make. It's stylish design prevents overhead glare and protects the front panel controls from accidental abuse. You'll find a use for this minibox design for all kinds of electronic goodies.

Figure 10-2 shows the plans. Transfer the dimensions of the minibox base to sheet metal. Section 10-1 offers some timely tips for working with sheet metal. It's best not to make the cover until the base is completed. Cut the metal with squaring shears, sheet metal shears, nibblers, or even a saber saw with a very fine-tooth blade. For easier cutting clamp the metal to 1/4" hardboard before saber sawing. Touch up any rough spots with a fine-toothed file. Sand the edges with fine-grit wet-dry abrasive paper. Drill holes and cut openings now before any bending takes place; it's faster and easier.

Bend flaps A, B, C, simultaneously at a 90° angle. Repeat with flaps D, E, and F. Then bend flap G and H. Now bend at line I 90°. Do the same at line J. Admire the minibox base; it's a beauty!

Before laying out the cowl cover, measure the outside dimensions of the completed minibox base. If the bends are accurate and square the measurements shown for the cover in Figure 10-2 will be satisfactory. However, sheet metal bending does not always turn out as planned so save yourself some grief and make a custom fit cover if the base dimensions turn out a little irregular.

Cut out the cowl cover, smooth up the edges if necessary, and bend the two sides 90°. Snap the cover over the base for a perfect fit. Clamp or tape the cover in place while lightly center punching holes at points marked X. Drill a 1/16″ hole through the cover and the base flap at each X point. After the pilot holes are drilled separate the two sections of the minibox. Redrill only the cover holes to 1/8″. Fasten the cowl cover to the base with #4 × 3/8″ sheet metal screws through the holes just drilled.

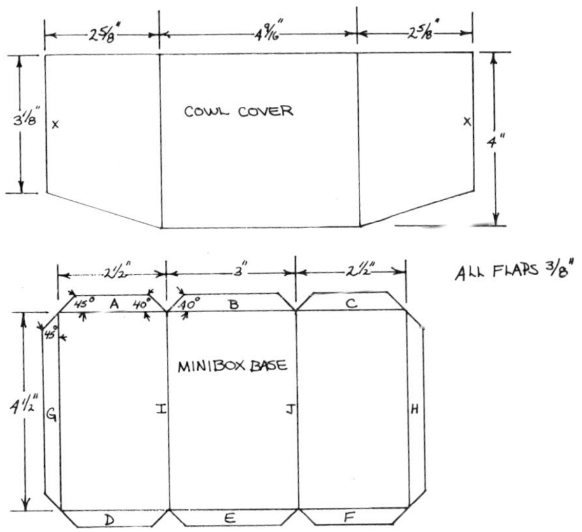

Figure 10-2. Cowl-type minibox plans.

If you want to make a cowl-type minibox with larger dimensions multiply each base dimension shown in Figure 10-2. For instance if you need a box twice as large simply multiply the dimensions by two (keep the flaps 3/8″). Once the base is completed custom fit the cowl cover to compensate for metal thickness and any bending irregularities.

10-3. BUILD AN ALL-PURPOSE CHASSIS

You'll like this chassis for it's clean and simple design. It is easy to make, has adequate inside space for parts and circuitry and has a touch of class, see Figure 10-3A.

The plans are shown in Figure 10-3B. Lay out the main section dimensions on sheet metal, preferably 18 gauge semihard aluminum. Refer to Section 10-1 for sheet metal tips.

Courtesy of Stenroos Electronics

Figure 10-3A. All-purpose chassis.

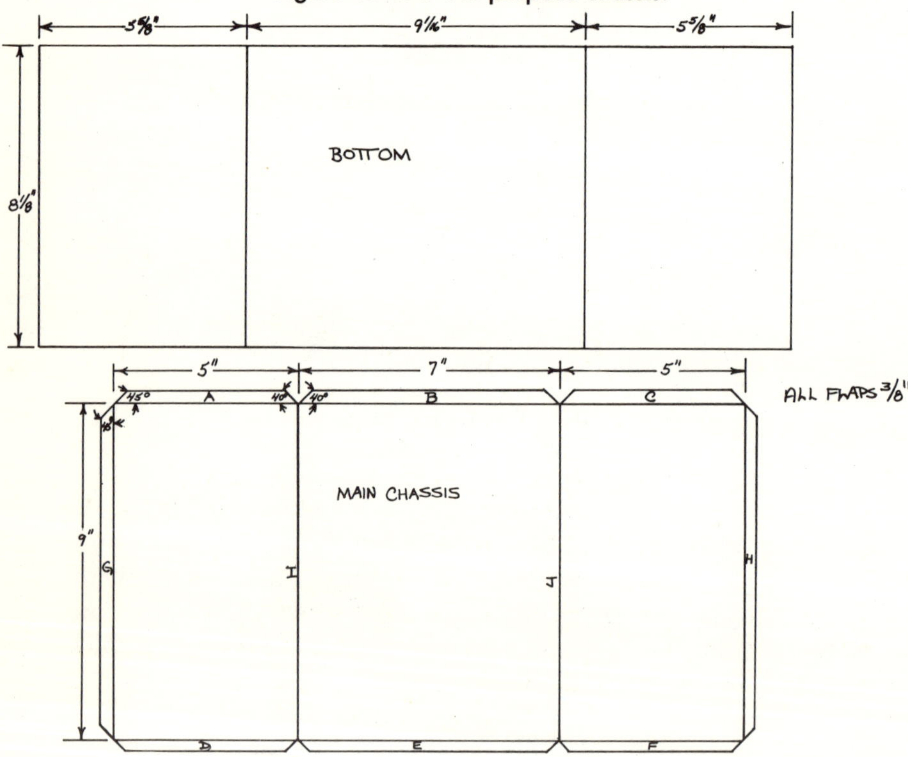

Figure 10-3B. Plans for all-purpose chassis.

Cut the pattern out using squaring shears, sheet metal shears, nibblers, or a saber saw with a fine-tooth blade. Next cut out the flap notches. Smooth up any irregularities or burrs with a smooth flat file and abrasive paper. Before bending, cut all holes and openings.

Bend flaps A, B, and C together at a right angle bend. Do the same with flaps D, E, and F. Finish the flaps with a 90° bend at G and then H. Next bend at line I, 90°. Repeat at line J. Inspect your work. Bend slightly here and there until the sides and bottom are square and perpendicular.

Before laying out the bottom section measure the outside dimensions of the main chassis. You may have to slightly change the measurements of the cover to compensate for bending irregularities or metal thickness. When you're sure of the exact size, scribe the bottom pattern on the sheet metal and cut out. Bend the sides perpendicular to the middle section. Smooth any rough edges.

Now for the big moment. Slide the bottom section over the main chassis . . . a perfect fit. For a super look, black vinyl the bottom section and spray paint the main chassis with a light color. Attach a rubber foot to each corner of the bottom. It's a beauty!.

10-4. BLUEPRINT FOR A STREAMLINED CONSOLE CABINET

This sharp looking chassis design, Figure 10-4A, is the perfect enclosure for that special test instrument project. Although the console cabinet appears to require sophisticated sheet metal work, it really is not at all difficult to make.

Courtesy of Calderone Electronics

Figure 10-4A. Streamlined console cabinet.

Start the project by scribing the layout lines for the console unit on sheet metal, see Figure 10-4B. Read Section 10-1 for timely tips on sheet metal fabrication techniques. Cut out the pattern with sheet metal shears, nibblers, squaring shears, or a saber saw with a fine-tooth blade. Notch the flaps as shown. Smooth up any rough edges with a fine-tooth file and abrasive paper. Before bending the console unit be sure to make all the holes and openings. It's much easier and quicker when the metal is still flat.

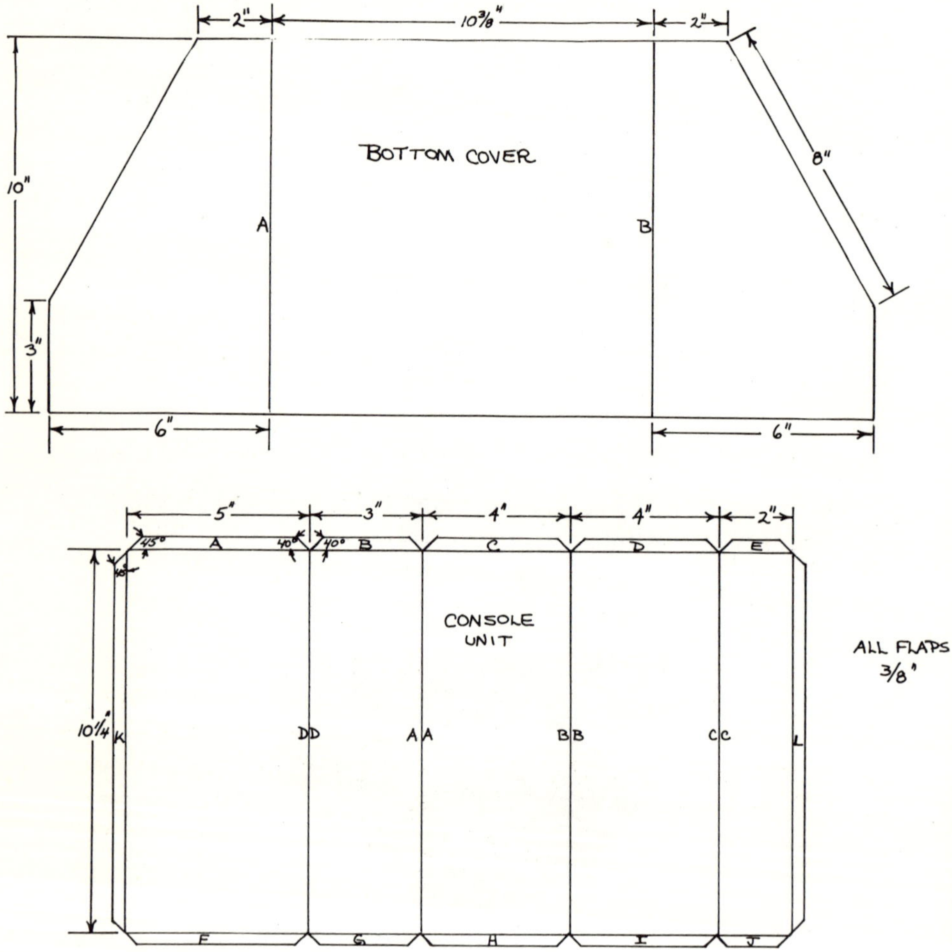

Figure 10-4B. Plans for a streamlined console cabinet.

First bend all the flaps 90°. Flaps A, B, C, D, E, can be bent together; so can flaps F, G, H, I, J. Next bend flap K and then flap L. Refer to Figure 10-4C while bending the rest of the console. Bend at line AA approximately 64°. Next bend at line

BB an angle of about 35° in the opposite direction. Now bend line CC about 75°. Finish with a 70° bend at line DD. Set the chassis in its normal position while inspecting. You'll probably have to bend a little more here and there until the chassis sits squarely on the workbench.

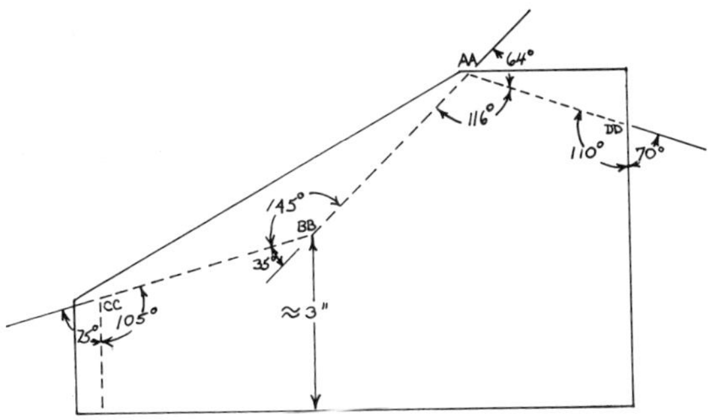

Figure 10-4C. Console unit bending angles.

Now measure the console to verify that its dimensions have not changed due to bending irregularities. Compensate for any measurement changes when laying out the cover pattern. When the cover pattern has been scribed, cut it out. Smooth up the edges. Bend at line A 90°, then at line B 90°. Square up the cover. Insert the console section into the cover . . . a masterpiece!

Finish up by fastening the two sections together with sheet metal screws. Lightly clamp or tape the two sections together before drilling. This will insure a perfect fit. Center punch and then drill a 1/16″ hole through the cover and the console unit at each fastening point. When these holes are completed, separate the two chassis sections and redrill only the cover holes to 1/8″. Use #4 × 3/8″ sheet metal screws to hold the sections together. Finish the console with a paint job, vinyl contact paper, or a combination of both. Attach a rubber foot to each bottom corner and the chassis job is finished.

10-5. A UNIQUE CHASSIS DESIGN

Are you looking for something different in an electronic apparatus enclosure? Maybe the model shown in Figure 10-5A is the answer. Here's a chassis that is different but still very utilitarian. It does not take a sheet metal expert to make a professional looking job. In fact the opposite is true. Just follow the easy directions for a neat and unusual chassis.

Courtesy of Misiak Electronics

Figure 10-5A. A unique chassis design.

Study the plans shown in Figure 10-5B. Scribe the pattern of the base unit into sheet metal; 18 gauge semihard aluminum is ideal. Refer to Section 10-1 if you have not read about tips for working sheet metal. Cut the pattern out of the metal with squaring shears, tinsnips, nibblers, or a saber saw with a fine-tooth blade. Finish the cutting by removing the notches from the flaps. File and sand any rough spots on the edges. Before bending the base unit be sure to make all the necessary holes and openings.

Individually bend flaps A, B, C, D, and E at 90° angles. Next make a right angle bend at line F; then at line G. Inspect, making slight adjustments to square up the base unit. Before laying out the cover section measure the outside dimensions of the completed base unit to be sure that the cover measurements are still meaningful. Irregular bends and metal thickness could change the cover dimensions slightly.

Lay out the cover pattern on the metal. Cut out as previously described. Finish up by cutting the notches from the flaps. Smooth up any rough edges. Bend the flaps H, I, and J together at a 90° angle. Next bend at line K so the side section is at about a 65° angle from the middle section. Repeat at line L. Try the cover on the base unit. You'll probably need to bend the sides a few more degrees to match up perfectly with the base unit. Don't overbend the cover angle; it's tough to correct. Save yourself some aggravation by bending the cover sides less than needed. Then gradually bend them to the exact size while custom fitting to the base unit. Slip the two chassis sections together . . . beautiful!

Spray paint and/or vinyl the enclosure, attach a rubber foot to each bottom corner and you're ready to label and then begin electronic construction. It's always a nice feeling to make a project that is a little different.

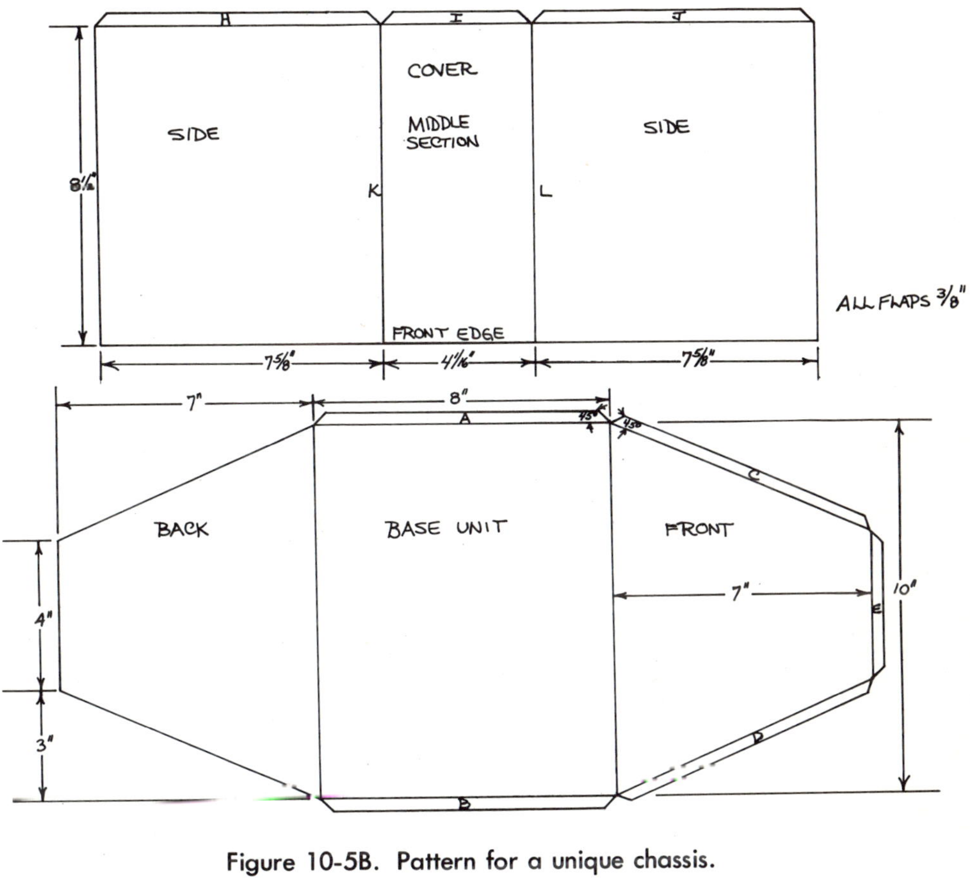

Figure 10-5B. Pattern for a unique chassis.

11 Practical Methods to Check Solid State Circuits

This section deals with straight forward effective methods to check various kinds of solid state circuits. You will read about testing transistor circuits with voltage and resistance measuring techniques. Typical SCR and triac circuits are described in relation to operation and troubleshooting. Practical information on field effect transistors, both JFET and MOSFET, is detailed.

Section 11-4 covers rectifier and zener diode circuits in a clear and concise manner. Many power supply problems can be solved by understanding rectifier and zener principles. You will also read about some down-to-earth integrated circuit service information. Neat tools and techniques for digital IC repair are described.

11-1. HOW TO MAKE VOLTAGE AND RESISTANCE CHECKS ON TRANSISTORS

A few quick DC voltage checks on transistor elements can often pinpoint a defective transistor or faulty circuit. There are two common methods to measure transistor voltages; element voltages in respect to ground and element voltages in respect to each other.

Most service information and schematics show transistor basing voltages in respect to ground. Figure 11-1A shows typical transistor voltage measurements. Many voltages listed on service data are with no-signal present; others list voltages with and without a signal. The AGC gate transistor shown in Figure 11-1A is an example of the two voltage type service information. The voltages shown in the triangles represent normal operation while the other voltages show no-signal condition. Be sure to identify whether the voltages are with or without a signal before troubleshooting. You probably will find this information in small print somewhere on the schematic. Also be sure to use a high impedance VTVM, DVM, or FET VOM to eliminate any voltage change caused by meter loading.

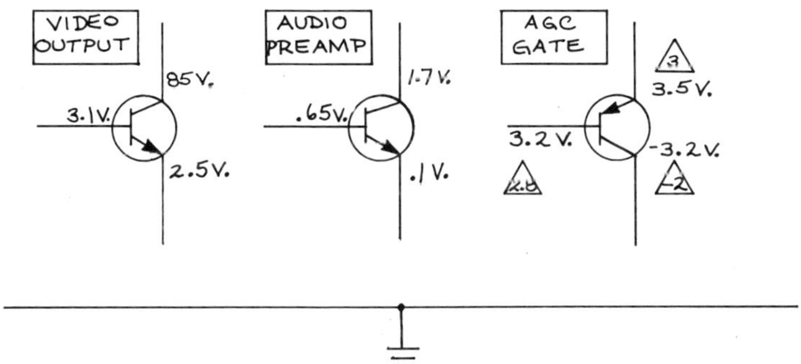

Figure 11-1A. Transistor voltage measurements in respect to ground.

Now let's voltage check the transistors in Figure 11-1A using the element to element method. Place the meter ground lead on the most negative of the two leads being checked. For example the voltage measured from the emitter to the base on the video output transistor would be .6V; the voltage between the emitter and collector would read 82.5V. The audio preamp transistor would check .55 volts between base and emitter. 1.6V would be measured between the emitter and the collector. No-signal emitter-base voltage of the AGC gate transistor would be .3V; emitter-collector would measure 6.7V.

The question remains what voltage measurement technique is the best. The answer . . . six of one, a half dozen of the other. Each method works as long as the meter is sensitive enough and the technician reads the meter correctly. It's easier to use the voltage to ground method when following a service schematic but small differences between elements are easier to read using the element to element method. An experienced technician will use both voltage measuring techniques depending on each individual situation.

Once erroneous voltages are detected it doesn't necessarily mean the transistor is bad. Checking the transistor, see Section 8-2, or substitution with a known good transistor will determine if the problem is the transistor or the circuit. If the circuit is bad resistance measurements can often find the defective culprit.

When making resistance measurements in solid state circuitry, an ohmmeter with low-ohms-voltage is necessary. Modern meters feature low-ohms function (less than .1V between probe tips) so the solid state junctions will not be turned on during the resistance measurements. Conventional ohmmeters will inadvertently bias junctions on, causing erroneous parallel resistance paths.

Figure 11-1B shows an audio module schematic. Suppose voltage measurements indicated a malfunction in the audio amplifier transistors. However both transistors check okay. Let's take some ohmmeter measurements. Remember the following points while ohmmeter checking with a low-ohms-voltage ohmmeter.

1. Most resistance readings are taken from elements of a tube, transistor, or IC in relation to ground.
2. Tubes, and solid state junctions can be considered open circuits.
3. Non-electrolytic capacitors block DC current and should be considered an open circuit.
4. Electrolytic capacitors have varying amounts of leakage resistance.
5. Resistance measurements for parallel resistor combinations will always be less than the smallest resistor.

Back to Figure 11-1B. Q 51 emitter reads 270 ohms which is normal. The base reads 64K ohms which also appears normal. The collector measures 100K ohms . . . definitely abnormal. Q 52 emitter and base both measure 100K ohms . . . strange. Q 52 collector measures 2700 ohms, okay. Abnormally high resistances found on Q 51 collector and Q 52 base and emitter are probably caused by the common power supply connection 12V (B). Further ohmmeter checking reveals a high resistance solder connection between the 12V (B) source and the PC foil. Reheating the connection and adding a bit more solder takes care of the problem. This particular problem was easy to diagnose with resistance measurements because the resistance paths to ground were fairly direct with minimum parallel paths.

Courtesy of Howard W. Sams & Co., Inc.

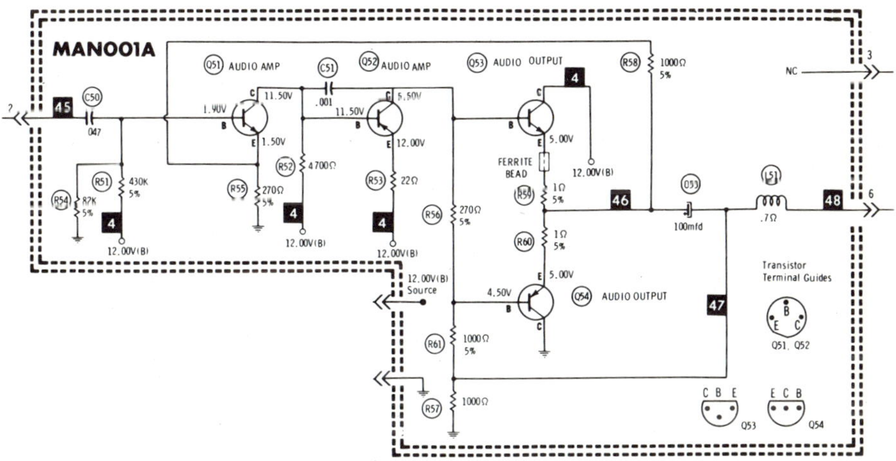

Figure 11-1B. Audio module schematic.

Good service information has resistance measurements already made and clearly listed. This info saves the technician a lot of time following and computing resistance paths from the schematic. Figure 11-1C shows a resistance measurement chart that really makes a breeze out of taking ohmmeter measurements. It would be nice if every schematic had a chart such as this to help the troubleshooter.

11-2. TESTING THYRISTORS AND ASSOCIATED CIRCUITS

The thyristor is a general term which defines a variety of solid state devices. Each of these devices can act like an electronically controlled switch with an on and off state. Most thyristors are used for controlling electrical power. Some common mem-

Courtesy of Howard W. Sams & Co., Inc.

RESISTANCE MEASUREMENTS

ITEM	PIN 1	PIN 2	PIN 3	PIN 4	PIN 5	PIN 6	PIN 7	PIN 8	PIN 9	PIN 10	PIN 11	PIN 12	PIN 13	PIN 14
V101	840 K	170 K	140 Ω	120 Ω	840 K	150 K	0 Ω							
MEASUREMENTS BELOW TAKEN WITH METER HAVING .08V MAX BETWEEN PROBE TIPS														
IC1	INF	INF	0 Ω	0 Ω	100 Ω	0 Ω	INF	INF	4000 Ω	4000 Ω	NC	4500 Ω	NC	350 K
ITEM	E	B	C		ITEM	E	B	C		ITEM	E	B	C	
MAF001D					**MAM001B**					**MAO002A**				
Q1	33 K	500 Ω	220 Ω▲		Q1	1300 Ω	6000 Ω	11 Meg		Q1	120 Ω	39 K	10 K	
Q2	470 Ω	1800 Ω	120 Ω		Q2	0 Ω	150 K	15 K		Q2	0 Ω	55 K	75 K	
Q3	180 Ω	1100 Ω	120 Ω		Q3	220 Ω	4000 Ω	850 Ω		Q3	0 Ω	85 K	3600 Ω	
Q4	3300 Ω	3200 Ω	0 Ω		Q4	100 Ω	3300 Ω	80 K		Q4	3600 Ω	100 Ω	85 K	
Q5	1200 Ω	7500 Ω	1200 Ω		**MAN001**					Q5	100 Ω	3600 Ω	470 Ω	
Q6	33 K	5000 Ω	270 Ω		Q51	270 Ω	65 K	4900 Ω		Q6	600 K	INF	120 Ω▲	
MAH002A					Q52	150-600 Ω	5000 Ω	2200 Ω		Q7	5000 Ω▲	5500 Ω	INF	
Q1	0 Ω	220 K▲	14 K		Q53	1200 Ω	2200 Ω	100 Ω▲						
Q2	0 Ω	150 K	470 Ω		Q54	1200 Ω	2000 Ω	0 Ω		**PW200**				
Q3	0 Ω	650 Ω	360 Ω		**CHASSIS**					Q1	8. 2 Ω	600 K	0 Ω	
					Q101	41 Ω	41 Ω	0 Ω		Q2	1 Ω	470 Ω	100 Ω	

▲ RESISTANCE MAY VARY.

Figure 11-1C. Resistance measurements chart.

bers of the thyristor family include the SCR, triac, diac, and the unijunction transistor.

In order for the SCR to operate, the anode must be positive in respect to the cathode as is the case for a silicon rectifier. However the SCR also needs a positive gate signal to start conducting, see Figure 11-2A. Once conduction begins, the gate loses control and only opening the anode or cathode circuit will stop conduction.

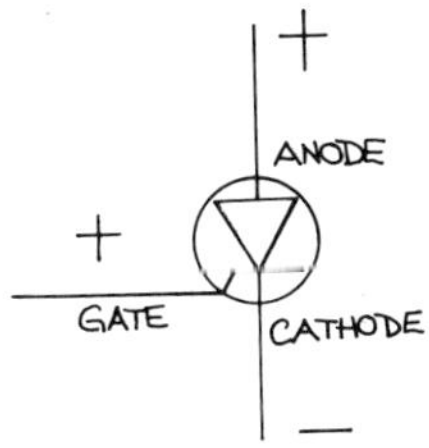

Figure 11-2A. Voltage polarity for SCR conduction.

Figure 11-2B shows a typical SCR regulated universal motor speed control such as would be found in portable electric tools and appliances. During the positive half cycle the gate receives voltage which is a combination of supply voltage and motor back EMF. The instant the gate triggers, in reference to the beginning of the positive half cycle of line voltages determines motor speed. This circuit has relatively constant speed characteristics under load since the back EMF decreases with load thereby allowing the SCR to fire earlier in the cycle supplying more power to the motor.

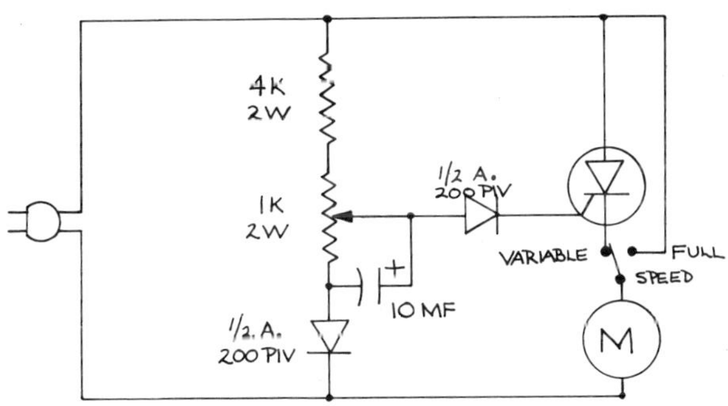

Figure 11-2B. SCR universal motor speed control.

Troubleshoot the SCR motor speed control circuit using the following techniques. If the full speed position works but not the variable speed the problem is probably the SCR or control circuitry. When neither variable or full speed positions work check the plug, line cord, switch, and motor windings. Very often the variable speed circuitry is built into the switch assembly making repair difficult. However as long as the assembly is not molded together you can usually find a way to obtain access to the individual parts. Section 8-2 and 8-3 describe testers that can make short work out of checking the SCR and the rectifiers.

The triac operates in a manner similar to two parallel SCRs (one inverted) with a common gate terminal. A triac will conduct with either a positive or negative potential on its anodes and gate, see Figure 11-2C. Since a triac is a bidirectional switching device it makes an excellent control mechanism for AC power. Refer to Sections 8-3 and 8-6 for information on a triac checker and a triac controlled AC power supply.

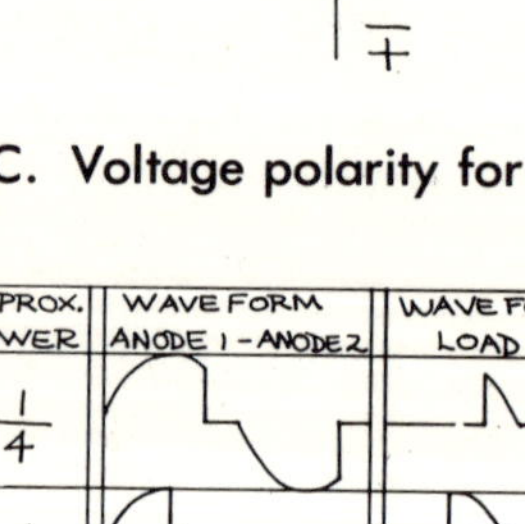

Figure 11-2C. Voltage polarity for triac operation.

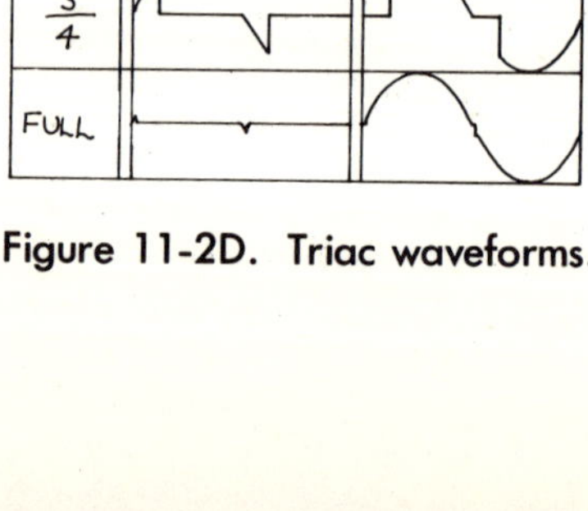

Figure 11-2D. Triac waveforms.

Figure 11-2D shows typical waveforms that are associated with triac controlled circuits similar to the AC power supply.

11-3. TIPS FOR CHECKING FETS

All FETs control current by means of a gate controlled electrostatic field. This phenomena results in a solid state device with an extremely high input impedance much like the characteristics of a triode vacuum tube. The three elements of a FET are called the gate, source, and drain. The input is the gate which is always reverse biased. The output is identified as the drain; the common element is the source. The two categories of FETs are the junction and the insulated gate (MOSFET).

The JFET symbols and equivalent resistances are shown in Figure 11-3A. An ohmmeter check between the source and drain would reveal a fixed resistance, 100 to 10,000 ohms, regardless of ohmmeter polarity. Measuring resistance from the gate to the other two elements would show high and low resistance depending on ohmmeter polarity and the gate diode. For instance if the negative ohmmeter prod is connected to the N channel source while the positive prod is on the gate, the ohmmeter should read the low forward resistance of the gate diode and the fixed resistance of the source. Reversing the ohmmeter prods would back bias the diode making the resistance reading extremely high. A P channel JFET could be ohmmeter checked in the same fashion; the only difference is that the gate diode is reversed.

Let's troubleshoot a JFET stage of a preamplifier circuit, Figure 11-3B. Suppose a signal tracer, refer to Section 12-3, reveals a good gate signal but no drain signal. Make a VTVM check between ground and the +35 supply; it's okay. Next measure the DC voltage across the 5.6K ohm source resistor. A voltage here would indicate FET channel current. Divide the voltage measured by 5.6K ohms to determine source to drain current. If no voltage is measured, VTVM across the source and drain. Supply voltage here would indicate an open channel or possibly a pinched off (cut-off) FET. Switch the input selector switch to an unused position, AUX or phono, to eliminate the possibility of the gate pinching off the channel. If no change is seen on the VTVM make an out-of-circuit resistance check on the FET. If the voltage drops when switching the input selector switch, check the tuner output circuitry. Similar voltage measurements can usually pinpoint all types of FET circuit malfunctions.

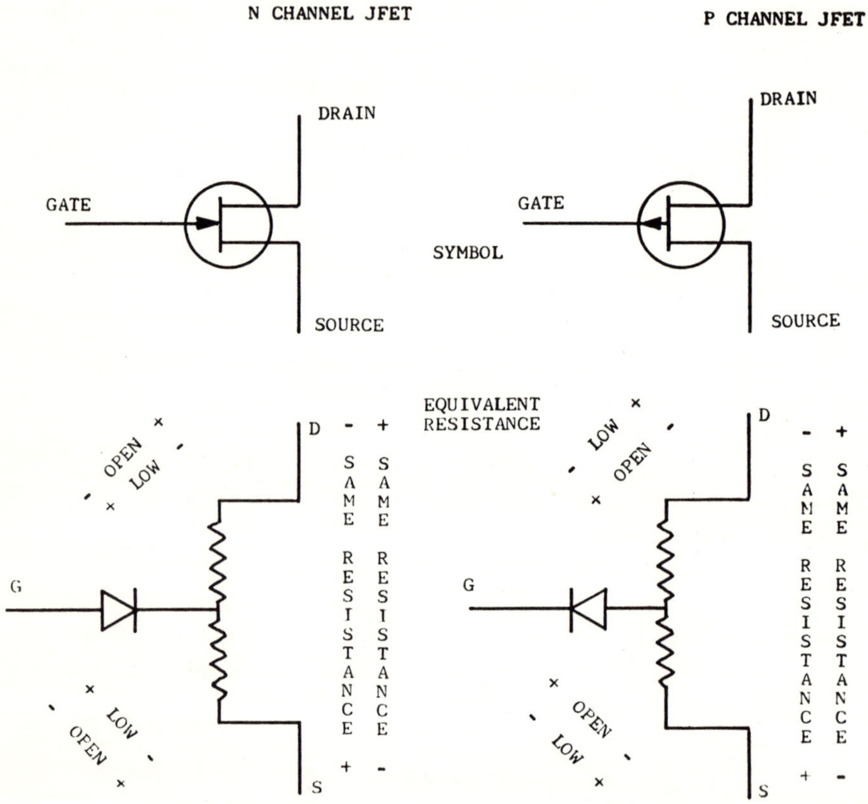

Figure 11-3A. JFET symbols and equivalent resistances.

Two types of MOSFETs are shown in Figure 11-3C. These FETs can be ohmmeter checked in a manner similar to JFETs. Some MOSFETs are very sensitive to static electricity and will actually be destroyed by handling. Static electricity can puncture the gate dielectric. When working with these devices grounding is the watch word. Keep your arms on a grounded metal plate when handling MOSFETs. Be sure to solder them using a three-wire iron with a grounded tip. MOSFETs of this

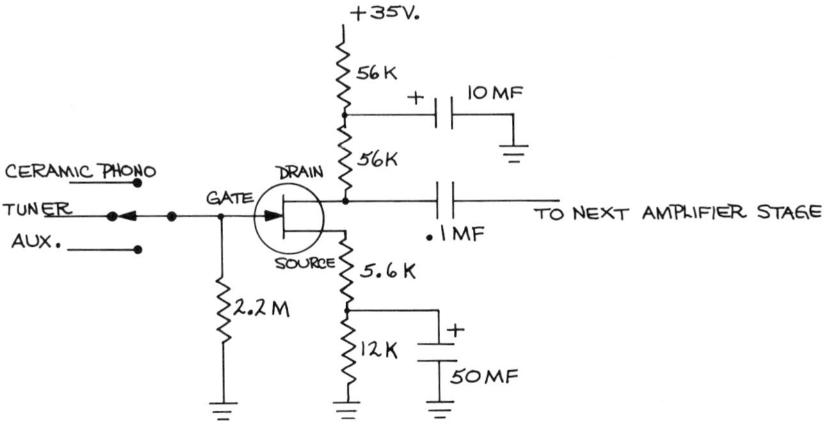

Figure 11-3B. JFET preamplifier circuit.

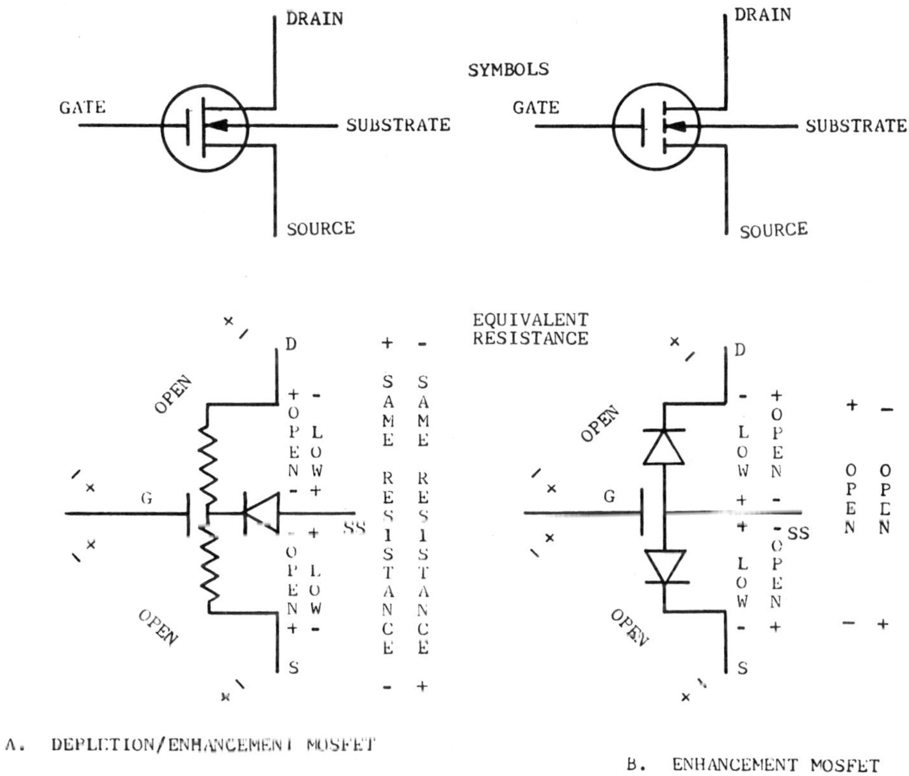

Figure 11-3C. MOSFET symbols and equivalent resistances.

nature are zot-proof packaged. A shorting ring around the electrodes or aluminum foil keeps them from being damaged by static electricity. Don't remove the zot-proofing until the MOSFET is soldered in the circuit. Modern MOSFETs are protected with gate zener diodes to eliminate excess gate voltage problems, see Figure 11-3D.

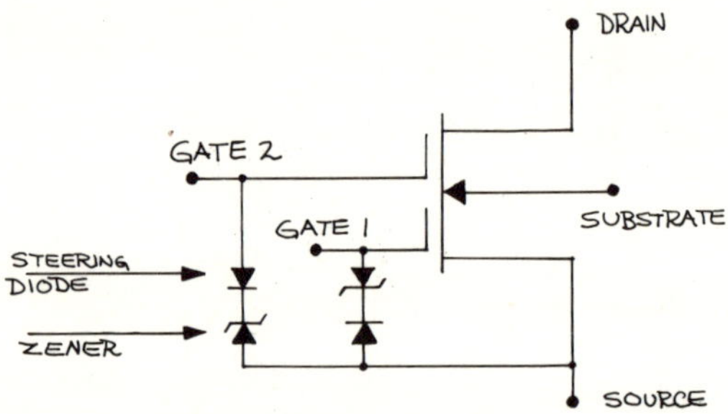

Figure 11-3D. Dual gate diode protected MOSFET.

11-4. IT'S EASY TO REPAIR RECTIFIER AND ZENER CIRCUITS

All electronic DC power supplies contain one type of rectifier or another to convert AC to DC. Most modern supplies also contain one or more zener diodes used for voltage reference. When troubleshooting any type of electronic equipment the power supply circuits are generally one of the first sections to check.

The DC voltage as well as the AC ripple voltage are the two standard measurements to make at the output of a power supply for a performance check. If either of these measurements deviate from the manufacturer's specs you are on the right track. A low DC voltage indicates power supply or load defects. Temporarily disconnecting the load will isolate the problem. This may mean slicing a printed circuit foil with a razor blade. If the DC voltage returns to normal the power supply is okay and a load circuit is at fault. When the voltage stays low the power supply is the problem. Excessive ripple voltage is almost always an indication of poor filtering in the power supply.

Silicon rectifiers in half-wave, full-wave, full-wave bridge, or voltage doublers form the common rectifying circuits, see

Figure 11-4A. AC voltage is supplied to the anode and positive DC voltage is available at the cathode in respect to circuit ground. Figure 11-4B shows oscilloscope waveforms for each of the common rectifier circuits. The top picture shows the rectifier(s) waveform under load with no filtering and the bottom picture shows the waveform with filtering. Notice how the capacitors tend to fill in the gaps and hollows between rectifier pulses. The difference between a full-wave and a full-wave

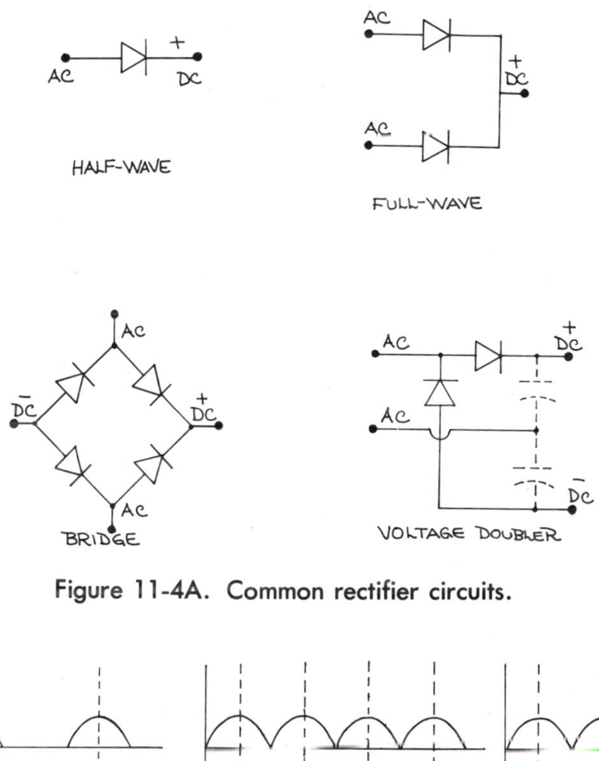

Figure 11-4A. Common rectifier circuits.

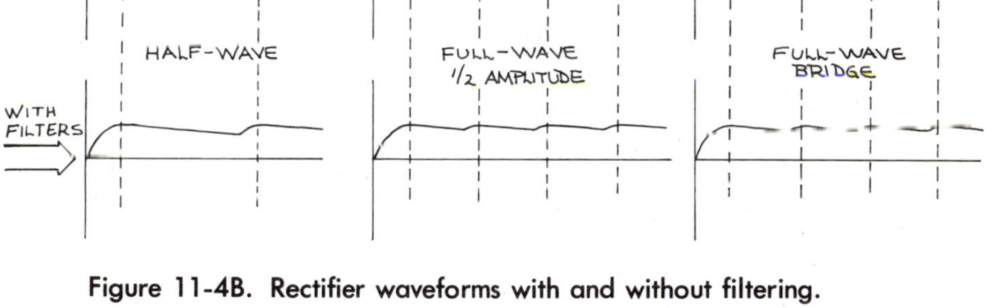

Figure 11-4B. Rectifier waveforms with and without filtering.

bridge rectifier circuit using the same supply voltage is that the bridge utilizes the entire supply voltage while the two rectifiers use one-half of the supply voltage.

To determine rectifier cathode DC voltage with a filter capacitor simply multiply the input RMS AC voltage by the multiplier shown:

RMS Voltage	Circuit	Multiplier	Cathode Voltage
6.3 VAC	Half-wave	1.4	8.8 VDC
6.3 VAC	Full-wave	.7	4.4 VDC
6.3 VAC	Bridge	1.4	8.8 VDC
6.3 VAC	Doubler	2.8	17.6 VDC

A silicon rectifier is easily checked with an ohmmeter. Disconnect one end from the circuit so there will be no erroneous parallel paths. The forward resistance should measure 20 to 50 ohms using the R x 1 scale of the ohmmeter. Reversing the ohmmeter leads will back bias the rectifier and show a high resistance meter reading; most silicon rectifiers exhibit back resistance characteristics of well over 100,000 ohms. In many situations when a rectifier is overloaded or shorts out it will self-destruct. The epoxy case will crack apart or simply disintegrate.

Figure 11-4C shows a partial power supply schematic using a zener diode for reference. Suppose you are having problems and you think the zener is bad. How do you check a zener diode? If the zener is working you should be able to measure the zener voltage across it with a voltmeter. For instance, in the circuit shown, a voltmeter check, positive prod on the cathode (top) and the negative prod on the anode (bottom), would reveal 20V if the zener was working. Of course for the measurement to be meaningful you must know what the zener voltage should be.

If the voltage check is incorrect an ohmmeter measurement should be tried. A zener will ohmmeter check exactly the same as a silicon rectifier if it's okay. To test for zener breakdown a test circuit must be constructed. Suppose you are checking a 20V, 1 W zener. It's a good idea to check the zener at 1/2 of its rated current. Use the following power formula:

$$P = I \times E$$
$$1\ W = I \times 20\ V$$
$$\frac{1}{20} = I$$
$$I = .05\ A \text{ or } 50\ MA$$

Construct the circuit shown in Figure 11-4D. The resistor value can be easily computed since 25 MA (1/2 zener circuit) should flow through it and 10 volts will be dropped across it. Ohm's law predicts the resistor will be 400 ohms. The resistor wattage using the power formula will be .25 W showing that a 1/2 W resistor will do the job. Voltmeter A will measure 10 V and voltmeter B will show 20 V if the zener is working.

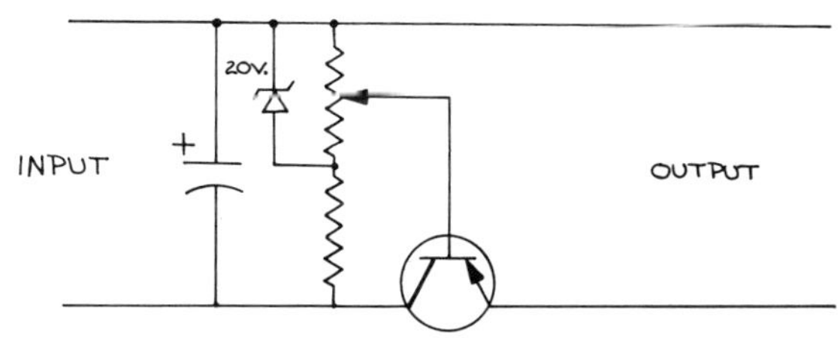

Figure 11-4C. Zener diode in a power supply circuit.

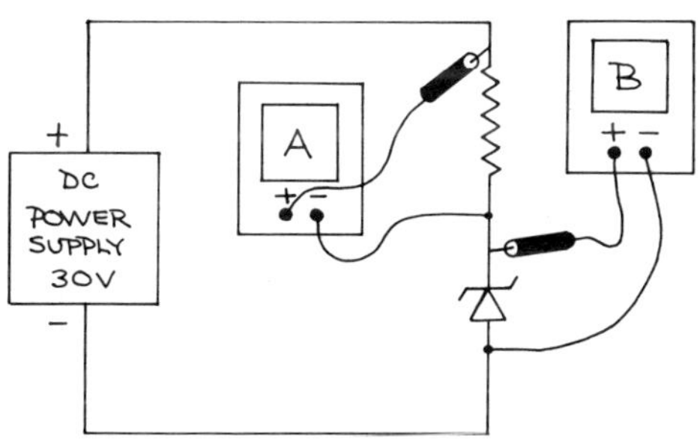

Figure 11-4D. Checking a zener diode.

11-5. POINTS TO CONSIDER WHEN TROUBLESHOOTING INTEGRATED CIRCUITS

Most ICs are enclosed in a round transistor-like TO-5 container or a dual-in-line package (DIP). The first step to any troubleshooting procedure should be to identify the pin numbers. Figure 11-5A shows the basing arrangement of the top and bottom views for the TO-5 and the DIP. Both these packages are similar to tube sockets in the manner that basing pins are counted. Count clockwise from the identifying mark when working from the bottom and counter-clockwise from the top side.

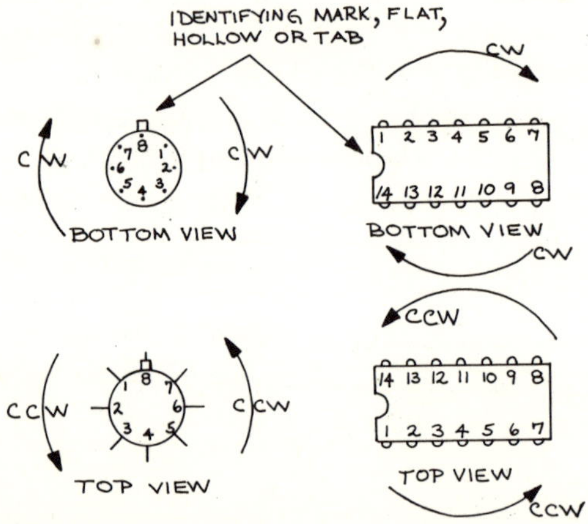

Figure 11-5A. IC basing arrangement.

Once the pins are identified be sure to check power supply voltages at the IC. Even the best IC can't work if the supply voltage is wrong or missing. You can't assume all power supply voltages are applied to the same terminals. The TO-5 type case usually has the positive power terminal connected to pin 8 and the negative to pin 4. The DIP configuration often has pin 14 connected to the positive and pin 7 to ground. Another common DIP combination is positive to pin 5 and ground to pin 10. The only sure way to identify the power terminals is with a schematic or the IC spec sheets.

There are two general classes of ICs, linear and digital. Linear ICs operate much like conventional electronic amplifying circuits but are packed into a much smaller area. Digital ICs operate using switching and counting principles, and are the prevalent type manufactured.

Of great interest to the technician are input and output voltages and waveforms. Whether the IC is a linear or digital type, if the input is correct and the output is incorrect or missing the problem centers around that particular IC. Careful voltage, resistance, and waveform checking at the IC pins will often lead to the trouble area. When all pin measurements seem to check out, the logical conclusion is that the IC must be defective.

Although the dual beam triggered scope is the most versatile test instrument for troubleshooting digital ICs, some special test apparatus is available that really is helpful. Figure 11-5B shows a logic clip that can be snapped over a 14 or 16 pin DIP. Each IC pin is monitored independently by an LED readout in the top of the clip. The technician can see at a glance the state of all IC pins; a high level signal will light the LED, a low

Courtesy of Hewlett Packard

Figure 11-5B. Logic clip.

level signal will keep the LED off. The clip uses no external power since power is drawn from the circuit. You don't have to be concerned about the IC basing in relation to the clip connections because the clip automatically adjusts for all pin configurations.

Another nifty tool for checking digital ICs is the logic pulser shown in Figure 11-5C. This probe has the ability to generate a logic LOW followed by a logic HIGH every time the trigger switch is pushed. The pulser will override any existing state that it is connected to for the duration of the pulse. Simply touch the pulser tip to an IC pin that you want to check. Push the button and a short pulse will be injected into the IC. If you are monitoring that particular IC with the logic clip, it's easy to see if the pulse is moving through the circuitry as it should.

Courtesy of Hewlett Packard

Figure 11-5C. Logic pulser

Figure 11-5D shows a logic probe that goes hand in hand with the logic pulser. This probe will read out by means of an integral light various circuit logic levels. For instance a high level signal, +2 V to +5 V, will cause the probe light to be bright. An open circuit or any voltage between +.8 V and +2 V will cause the lamp to dim. A low level signal, 0 V to +.8 V will extinguish the light. Pulses greater than 10 nanoseconds will cause the probe light to display a .05 second blink. The light blinks on or off depending on the pulse polarity. Pulse trains (repetitive pulses) up to 50 MHZ will cause the probe lamp to blink at a 10 HZ rate.

Courtesy of Hewlett Packard

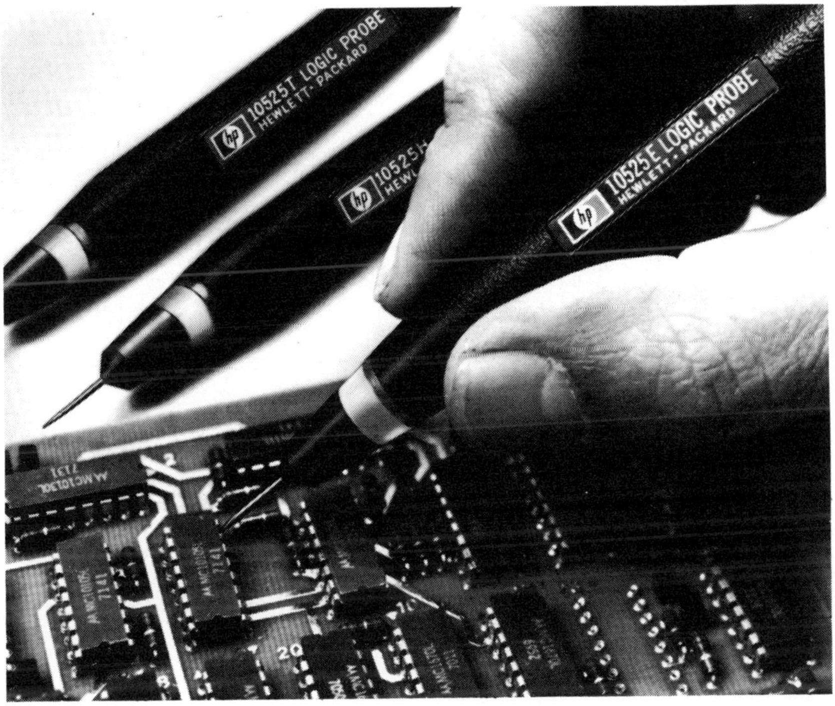

Figure 11-5D. Logic probe.

Let's demonstrate how the logic pulser and logic probe can be used together to check out a typical TTL (transistor-transistor-logic) IC. Figure 11-5E shows a 7432 quad 2-input OR gate. First check the +5 V at pin 14 with the logic probe. The probe light should be bright. Checking pin 7, ground, should cause no light on the probe. The 7432 IC contains 4 OR gate circuits which could be used independently or interconnected. For our discussion let's say they are connected separately. Here are the rules for OR gates:

1. Any high level input will cause a high level output.
2. A low level output can only result with all inputs low level.

Now inject a pulse into pin 9 (input) of one OR gate while monitoring pin 8 (output) with the logic probe. If this part of the IC is working, a blink will be seen on the probe. The same effect should be noted when injecting a pulse into pin 10. The other three OR gates in the 7432 can be checked in the same manner.

The combination of the logic clip, logic pulser, logic probe, and the basic knowledge of digital circuit rules is all that's necessary to effectively troubleshoot most digital equipment. No matter how complicated or sophisticated digital equipment appears, the basic circuitry always boils down to small logic blocks like OR gates that operate according to very simple rules.

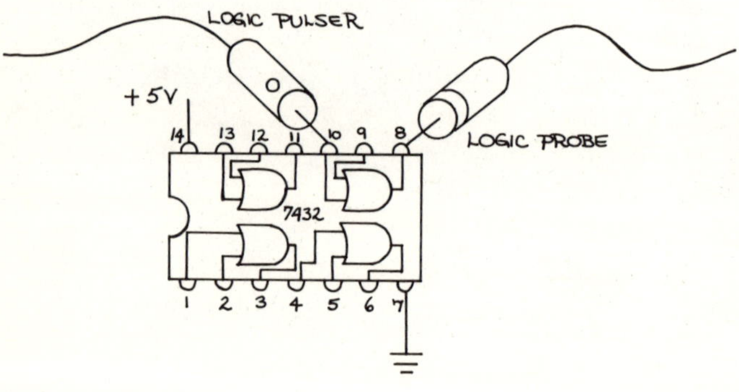

Figure 11-5E. Checking a TTL 7432 IC with a logic pulser and logic probe.

12 Easy-to-Use Specialized Electronic Test Equipment

Five worthwhile electronic test instruments will be discussed in this section. Each instrument, although specialized in its function, is an important addition to any electronic workshop.

Read about effective techniques for checking picture tube circuits. You'll find out how to diagnose a bad picture tube and how to electrically rebuild a picture tube without removing it from the cabinet. Section 12-2 discusses practical uses and features of the digital frequency counter; definitely an instrument that clears up a lot of calibration bug-a-boos.

Stereo tape player problems become minimal when you're troubleshooting with a signal tracer as shown in Section 12-3. Audio problems practically fix themselves when you know how to use a signal tracer. Methods and techniques utilizing a sweep generator are covered that make short work of electronic alignment problems.

In the last section white dot and color bar generator patterns are discussed. You'll find just how indispensable this instrument can be for color television servicing.

12-1. HOW TO EFFECTIVELY TEST PICTURE TUBES

Picture tubes are expensive! The most costly single part in any television set is the picture tube (CRT). It is also a part that eventually will go bad and have to be replaced. However before a picture tube replacement is considered, careful systematic evaluation and measuring techniques are required. It's a very embarrassing situation (and costly) when a color picture tube has just been replaced and the TV still has the same trouble symptoms.

A wise technician will always check the power and signal inputs to the CRT when a bad tube is suspected. If all of these measurements check okay, the external circuitry can be eliminated as the trouble source. Let's analyse a typical repair job where the picture tube is suspected of being defective.

Figure 12-1A shows the schematic and service information for a color CRT. A visual inspection will show if all heaters are lighting. If one heater is out and the connections to it measure 6.3 VAC you can be sure that the CRT heater is open. Make it a policy to always check heater voltage when the heater(s) are dark. The chances are the heater has burned out but you can't be sure until you've checked the incoming voltage.

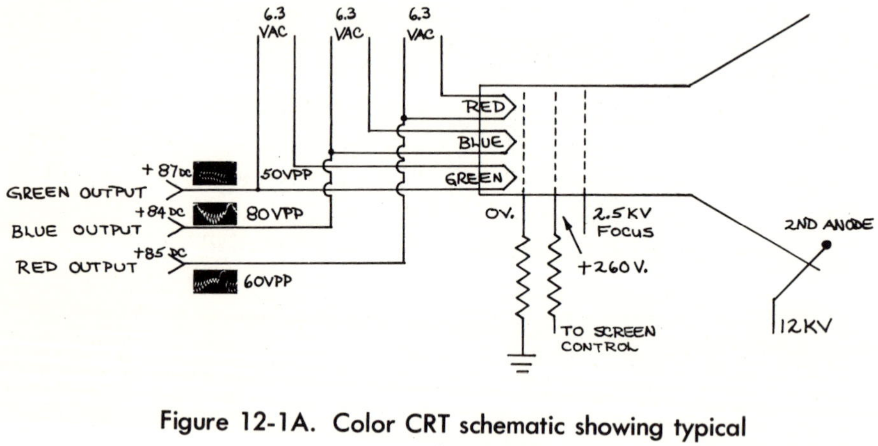

Figure 12-1A. Color CRT schematic showing typical service information.

Next check all DC voltages at the picture tube socket. Generally it's possible to slip the socket backwards slightly in order to make the CRT pins accessible for testing. Compare the readings with the data on the service information. Be sure to use a DC high voltage probe when measuring the focus voltage (refer to Section 7-3) and the high voltage on the second anode. The safe way to measure the CRT high voltage anode is as follows:

1. Turn off TV power.
2. Connect meter ground to HV ground.
3. Slip high voltage probe tip under second anode boot. Many HV probes have a wire attachment designed to hook the tip on the HV cavity. Don't touch with fingers since an HV charge may be present.
4. Double check the range and function switches of your meter.
5. Turn on power and read HV on the meter.

If all the previous measurements have checked out, scope the video signals into the picture tube. Figure 12-1A shows waveforms and P to P voltages taken with a color bar signal (see Section 12-5) injected into the antenna terminals. You're one step closer to confirming a bad CRT if the video waveforms are also correct.

Now's the time to use a picture tube checker to find out exactly what's wrong with the CRT, see Figure 12-1B. It's also a reassuring confirmation that the picture tube is really bad. A good CRT tester will test for shorts between the heater(s) and cathode(s); check for leakage on the control grid(s). It will measure emission or beam current for each gun. You'll be able to check for proper tracking between guns in color tubes. Most checkers have a provision for predicting the remaining useful life of the picture tube. This is a very handy feature to have when contemplating replacing a marginal tube.

In addition to checking CRT's, a good picture tube checker can actually repair many CRT defects. Shorts associated with the control grid can often be removed by a special short removal function. Low emission on a single gun, two guns, or even all three guns can usually be corrected with rejuvenation procedures outlined in the picture tube tester instruction booklet. Although rejuvenation of a CRT is never 100% assured, you'll be surprised how often this simple procedure is effective. It's common for a rejuvenated picture tube to last for a year or more.

Courtesy of Sencore

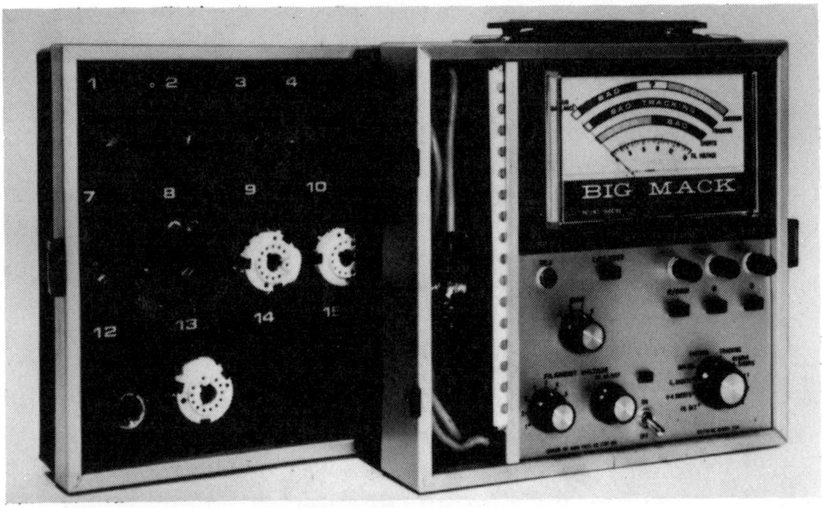

Figure 12-1B. CRT tester.

If voltage measurements, waveform analysis, and a CRT checker confirm that a picture tube is bad and can't be rejuvenated, you can rest assured that the tube needs to be replaced.

12-2. WORKING WITH THE DIGITAL FREQUENCY COUNTER

The complete electronic workshop should not be without a high-speed multifunction digital frequency counter. It was not too many years ago when ownership of a frequency counter was just out of the question financially for the average technician or electronic enthusiast. Digital ICs have changed the rules and made it possible to have a fine electronic counter at a reasonable price. Built it yourself from a kit and the price is even less, see Figure 12-2A.

How many times have you wondered about the frequency accuracy of your generators and other precision test instruments? Especially after a few years of everyday use. You know the equipment needs checking and recalibration but what are the options?

1. Ignore the problem and work with substandard equipment.

2. Pack and ship the equipment back to the factory . . . a hassle.
3. Buy a new instrument . . . just for the accuracy.

The chances are you'll try to live with what you have until you're fed up with the problems created by inaccurate test equipment. Then on the spur of the moment you'll order a new instrument(s) to replace a perfectly good one that perhaps just needs recalibration.

Courtesy of Heath Co.

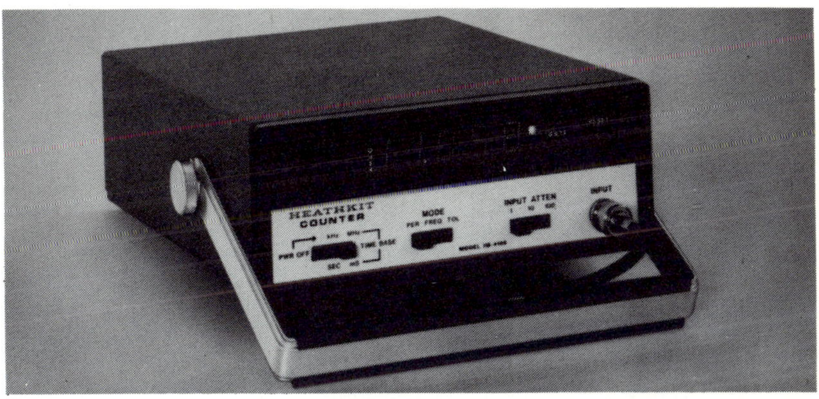

Figure 12-2A. Digital frequency counter kit.

Figure 12-2B shows how a digital frequency counter can be used to check an RF generator for frequency accuracy. Start out on the lowest frequency band of the generator. Set the counter for the appropriate frequency range. Turn the generator dial through the frequency range while monitoring the digital frequency counter. Any frequency errors in the generator will be obvious. It's a good policy to jot the types of errors found on each generator range. This information will be most helpful when recalibrating the various ranges. How many generators have been back-shelved just because a dial knob slipped or one range deteriorated? This simple check only takes a few moments to clear any frequency uncertainties.

Many digital frequency counters can be set to count and totalize external events. This simply means that you can count low frequency cycles or pulses in a nonrepetitious pulse train for a given amount of time. An overflow indicator on most counters shows when the count has exceeded the capacity of the instrument. Pushing a reset button is generally all that's necessary to clear the counter in order to start a new sequence. It's not an unusual capability for counters to totalize up to 999,999 events.

Figure 12-2B. Checking accuracy of an RF generator with a digital frequency counter.

Another nice feature of some counters is the digital tachometer readout of revolutions per minute (RPM). The counter is connected to a pulse per revolution transducer for the RPM reading. Counters are usually calibrated to work with transducers that produce 100 pulses per revolution. Other transducers will also work if the following formula is used:

$$\text{RPM} = \text{readout} \; \frac{100}{\text{pulses per revolution}}$$

For most general electronic related applications, counters are not used nearly as much as other test equipment. However, when the need for a frequency counter arises (it will be more often than you think) it surely is a pleasure to have one readily available.

12-3. MASTER AUDIO DEFECTS TROUBLESHOOTING WITH A SIGNAL TRACER

A signal tracer is just the ticket for mastering bugs in radio and audio equipment. It's sweet and simple to use and will locate a lost signal quicker than you can say Jack Robinson. The tracer itself is nothing but a very high gain audio amplifier with the additional capability of detecting an RF signal. Many tracers have a switch in their probe to control operation for use in an audio or IF-RF circuit. You don't have to be concerned with damaging even the most sensitive circuit while signal tracing since the tracer applies no energy to the circuit and only uses the signal that is already present.

Let's troubleshoot a stereo tape player, Figure 12-3, using signal tracing techniques. The right channel is working perfectly and the left one is dead. We will assume the power supply is okay since both channels operate with a common supply. We have already tried several cassettes so a bad tape is ruled out. Cleaning the head was also tried to no avail. Turning up the left channel volume control produces hum in the left speaker indicating that the circuit from the volume control to the speaker is probably okay.

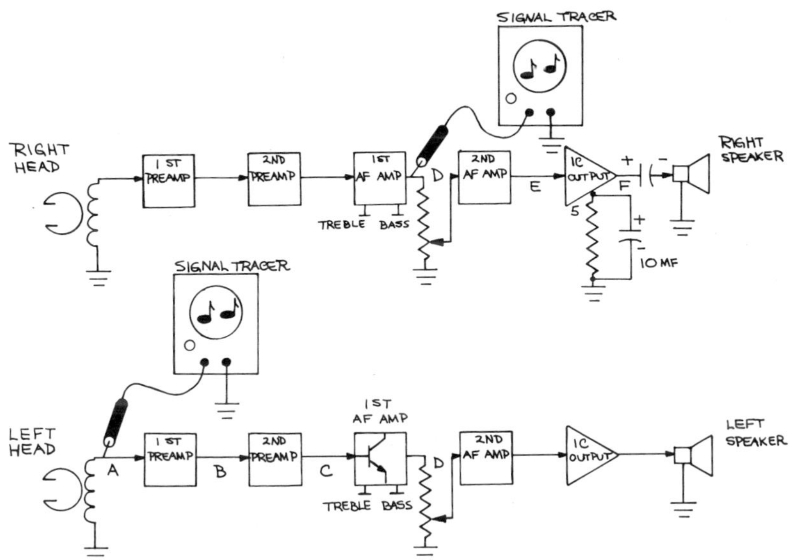

Figure 12-3. Stereo tape player block diagram.

Put a cassette in the player and fire up the signal tracer: it's time to find that electronic bug. With the tracer's gain control turned up sweet music is heard from the signal tracer's speaker as test point A is probed, see Figure 12-3. Next probe test point B. The music is louder indicating amplification is occurring in the 1st preamp stage . . . okay. Move the probe to point C. Mmmm, more music with a louder volume; the second preamp must also be working properly. Probe at point D . . . nothing. the bad area is between the second preamp output and the first

AF amplifier output. The tracer follows the signal right to the base of the first AF transistor but no sound is heard on the collector.

A voltage check reveals +14 V on the collector and 0 V on the emitter. The service info shows the collector sitting at +6 V and the emitter at .5 V. Looks like an open transistor. Pop the 1st AF AMP transistor out and check it, refer to Section 8-2. It's bad. A new replacement transistor is soldered in the circuit; lo and behold the left channel is back in business.

Here's another case. This time the left channel is okay but the right channel is distorted especially on high volume. Trying a different tape and cleaning the heads does nothing for the problem. Maybe it's the speaker; interchange the left speaker with the right. The problem is still the same. We are ready for the signal tracer. With a tape playing, signal trace at the top of the volume control, test point D in Figure 12-3. Good sound here with no distortion. Next probe at the second AF output, testpoint E. No distortion is heard from the signal tracer even at high volume. Probing at testpoint F, the right channel IC output reveals distortion—the same kind as heard from the tape player's speaker.

Measuring voltages on the IC pins reveals no really meaningful clues except that pin 5 is slightly above normal. It could be the output IC but there are two electrolytic capacitors that must be suspected first. Electrolytic capacitors are noted for causing problems. Besides, it's a lot easier to check or substitute a capacitor than a 10 pin IC. Replacing the 10 MF bypass capacitor with a new one clears up the distortion problem.

As shown in the two previous examples the signal tracer coupled with fundamental troubleshooting procedures makes short work of localizing troubles to a particular stage. Once the bad stage has been identified, voltage and resistance checks will usually pin down the defective part(s).

12-4. PRACTICAL WAY OF UTILIZING A SWEEP GENERATOR

A sweep generator is a special RF generator that is frequency modulated. Instead of producing a single frequency RF signal, it will produce a band of frequencies that sweep back and forth over a selected range. A sweep signal is necessary for evaluating, aligning and adjusting TV, FM, and other circuits which operate over a wide bandwidth. A sweep generator along with an oscilloscope enables the technician to actually see the overall response curve of a stage or section.

When using a sweep generator, a conventional RF generator is also necessary to mark various frequency points on the response curve. Since the response curve is the result of many frequencies, the only way to identify individual frequency points along the curve is to mark them with the RF generator. Figure 12-4A shows a typical TV video IF response curve. Note how various key frequency points along the curve are identified

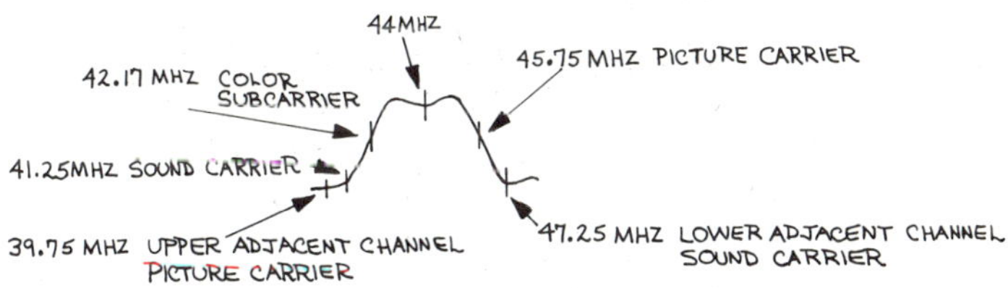

Figure 12-4A. Typical TV video IF response curve.

with marker pips. As the technician changes the frequency of the marker RF generator, the pip will move along the response curve. If the technician sets the generator at 44 MHZ, then the pip should fall in the top middle of the response curve (providing the IF strip is aligned and operating properly). Many modern sweep generators have built in marker frequencies, Figure 12-4B, which eliminates the use of a separate marker generator. They really simplify and speed up the alignment process.

Let's make a TV chroma bandpass alignment using a sweep generator (built in marker), and oscilloscope. The equipment is connected as shown in Figure 12-4C. Always heed the manufacturer's instructions when making an alignment or response check. Special instructions pertaining to TV controls, sweep synchronization, AGC bias, clip lead placement, etc. should be followed before attempting any evaluation or alignment adjustments. Now all the controls are set properly and the preliminary adjustments finished. The apparatus is all connected and ready to be turned on. Apply power and monitor the scope. Adjust both first and second chroma bandpass amplifier transformers for maximum gain and symmetry as shown in Figure 12-4D. This completes the first step in the chroma alignment.

Courtesy of Sencore

Figure 12-4B. Sweep and marker generator.

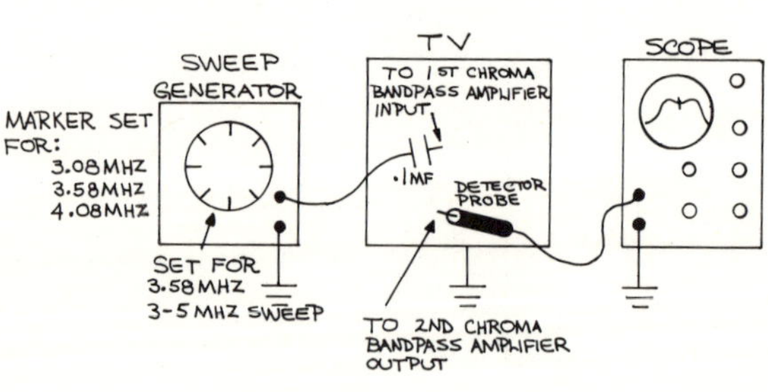

Figure 14-4C. Equipment set up for a TV chroma bandpass alignment.

Figure 14-4D. Chroma bandpass amplifier response curve.

Now connect the sweep generator output to the VHF tuner test point through a .001 MF capacitor. Change the sweep frequency to 44 MHZ with a 10 MHZ sweep. Adjust the 1st video amplifier transformer for maximum gain and symmetry as shown in Figure 12-4E. Touch up the chroma bandpass amplifier transformers slightly. Adjust the 4.5 MHZ trap for minimum as shown in Figure 12-4E. This completes the chroma bandpass alignment.

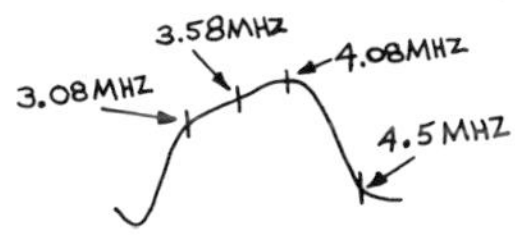

Figure 12-4E. Video and chroma bandpass amplifier response curve.

It's much more involved to read about an alignment procedure than to actually do it. Once the equipment (sweep generator, scope, etc.) is familiar to the technician and the various test points and transformer locations are physically identified, the alignment procedure actually takes only a few minutes. It's the kind of thing that seems to take forever the first time, but when familiar with the procedure it's quite fast and easy.

12-5. POINTERS FOR WORKING WITH A WHITE DOT AND COLOR BAR GENERATOR

The white dot and color bar generator is a must for working with color TV's and is also useful for black and white sets Most generators produce the TV screen patterns shown in Figure 12-5A. The pattern signal can usually be injected into the TV antenna terminals riding on an RF carrier or be directly inserted into the video sections.

The vertical, horizontal, and crosshatch patterns are used for checking and adjusting linearity controls. Crosshatch and dots are used for making static and dynamic convergence adjustments on the color CRT. The color bar pattern is especially well suited for signal injection and troubleshooting. Shading bars are used for checking B & W tracking.

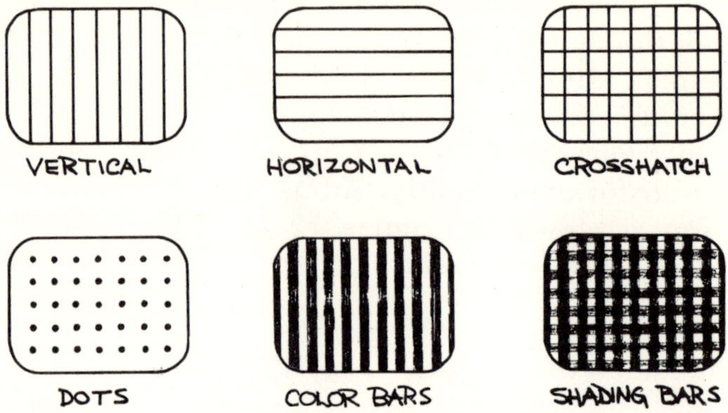

Figure 12-5A. TV screen patterns produced by a white dot and color bar generator.

Let's check linearity in a typical TV. Set the TV channel selector knob to match the frequency of the bar generator; a channel that is not being transmitted in your particular area. Connect the RF generator output to the TV antenna terminals. Switch on a vertical bar pattern. The bars should be evenly distributed, the same thickness, and straight up and down with no bending. TV controls that have the most effect on this pattern are the horizontal linearity and width controls.

Now switch the generator to horizontal bars. Like the vertical bars they should be evenly distributed and of equal thickness. The bars should be parallel with each other over the entire screen width. TV controls to adjust are the vertical size, height, and vertical linearity.

The crosshatch pattern is used to check for vertical and horizontal linearity simultaneously. Ideally the pattern will show equal size squares over the entire screen. The vertical and horizontal controls may have to be touched up slightly to compensate for any irregularities. It's also a good procedure to switch the TV channel selector back to a live channel to make sure the picture fills the screen normally. It's possible to have a perfect crosshatch pattern yet have the TV picture way out of proportion.

Dots and crosshatch patterns are just the ticket when checking or adjusting the convergence controls. A finer dot or bar pattern can usually be obtained with video injection rather than RF. If you're just checking convergence use the RF injection; for actual convergence adjusting use video injection. When viewing the dot pattern on the color TV screen each dot should appear white. The dots will appear white when the red, green, and blue colors are perfectly converged. Technicians usually use the dot pattern while adjusting the center area static convergence adjustments. The border areas which are converged with the dynamic convergence controls are generally easier to make with a crosshatch pattern. The interaction of the adjustments can be minimized by carefully following the manufacturer's convergence instructions.

Use the color bar pattern for checking the operation of the color circuits. The color intensity of the bars should be adjustable with the TV color control. The TV hue or tint control should be able to change the hues of the color bar pattern.

Inject a color bar pattern into the antenna terminals when troubleshooting color circuits. The fixed pattern makes an ideal signal to trace using an oscilloscope. Most service information shows key waveforms taken with a color bar generator signal present. Common bar pattern waveforms are shown in Figure 12-5B. It's much easier to compare scope waveforms with service info waveforms for shape, amplitude, and frequency when a common signal is being received.

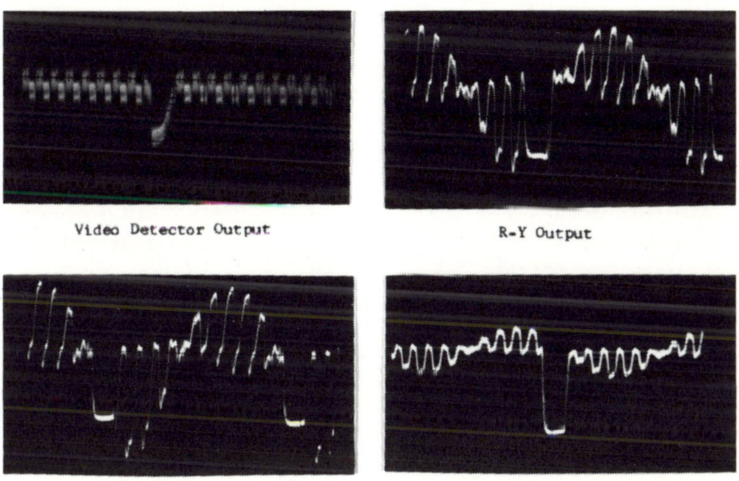

Video Detector Output — R-Y Output

B-Y Output — G-Y Output

Figure 12-5B. Video waveforms with color bar signal.

The shading bar pattern is great for making black and white tracking adjustments. The screen and gain controls are adjusted to produce good black, white, and gray pictures throughout the range of the brightness and contrast controls. These adjustments are difficult to make when using a constantly changing picture from a broadcasted signal. The stationary shading bar pattern usually consists of a wide bar cross-hatch pattern that produces four distinct levels of brightness, see Figure 12-5A.

13 Guide to Understanding Oscilloscope Waveforms

What electronic workshop would be complete without the most versatile instrument of all . . . the oscilloscope. This section is devoted entirely to this magical electronic machine.

Read about a triggered dual trace scope. Setting up the controls is easier than you might think. You'll love a dual trace scope; it's twice as good as a one beamer! Find out how square waves can be used to solve audio dilemmas. A square wave analysis can tame the meanest audio bug.

Section 13-3 gives pointers for using Lissajous patterns. Learn how to convert your scope into a vectorscope for color TV work. The composite video waveform is chock full of clues and info for the TV troubleshooter. Read about how the composite video waveform helps to solve your bug-a-boos.

13-1. SETTING CONTROLS OF A DUAL TRACE TRIGGERED SWEEP SCOPE

In this section we'll emphasize the controls that are set and read differently than the usual general purpose scope controls. Figure 13-1A shows a dual trace oscilloscope with triggered sweep. Refer to this picture for clarification as you read the following material. Although all scopes are not marked or made exactly the same, once you become familiar with any dual trace triggered sweep scope, you'll have little difficulty operating any model.

Photo courtesy of Tektronix, Inc.

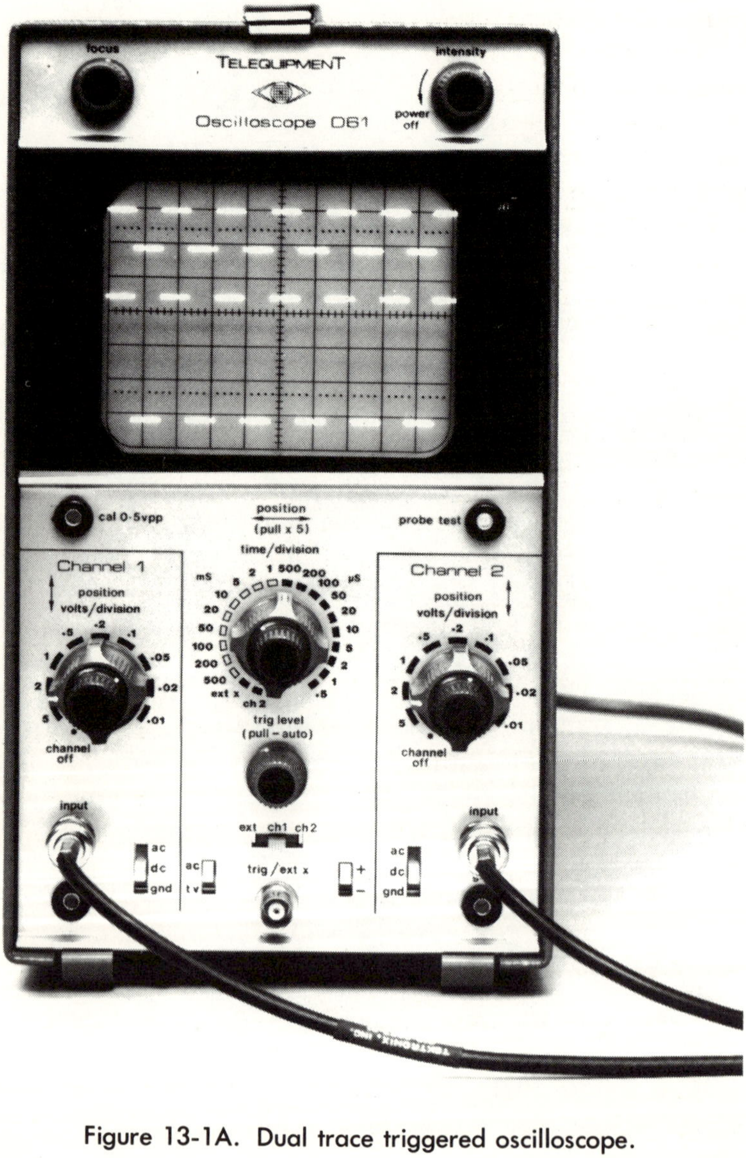

Figure 13-1A. Dual trace triggered oscilloscope.

Notice how channel 1 and channel 2 controls are separated on the right and the left side of the front panel. Controls that are common to both channels are located in the middle area and top of the front panel. Channel 1's inside knob controls the vertical position of the beam as well as turning the trace off in the extreme CCW position. The outside knob for channel 1 controls the vertical sensitivity or volts/division.

The plastic graph, graticule, over the CRT face is marked in square centimeters (cm.). There are 10 cm. of display area in the horizontal direction and 8 cm. in the vertical. Suppose channel 1's volt/division switch is in the .1 position and the waveform appears as shown in Figure 13-1B. Since the sawtooth amplitude fills 2 cm. then its peak to peak voltage will be 2 × .1 or .2V PP. Simply multiply the volt/division switch setting by the number of cm. that the waveform fills in the vertical direction.

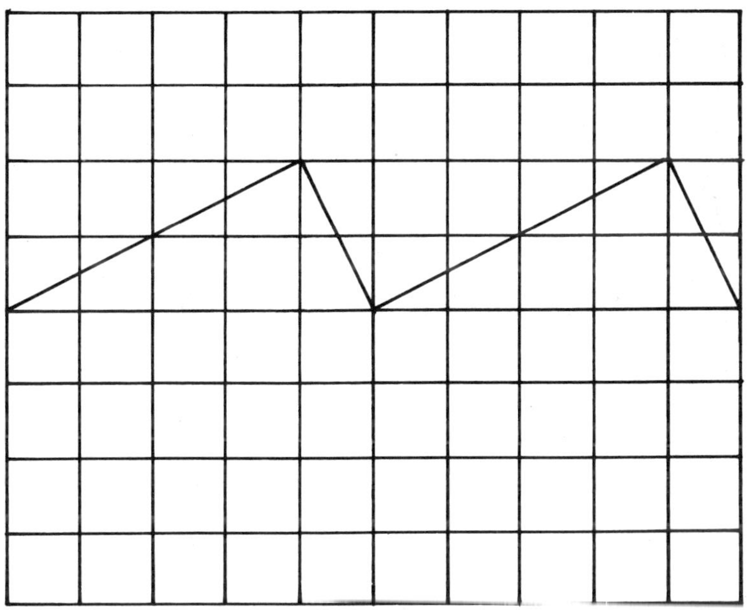

Figure 13-1B. Oscilloscope sawtooth waveform display.

The ac, dc, gnd slide switch in the lower part of channel 1's area selects the input signal coupling. The gnd position automatically opens the external input connector and grounds the internal scope amplifier. This feature is great for determining the exact zero line for a trace. The dc position couples the input directly into the scope. The display will show the waveform plus its DC level. The ac position removes the input signal's DC component only allowing the AC part of the signal to be displayed. For instance, in Figure 13-1B the sawtooth waveform would drop and center itself around the zero line; the DC component is ignored. The zero line does not have to be in the middle but can be set anywhere by the operator just by switching to the gnd position and turning the position control.

Locate the trigger level switch in the lower middle of the front panel. Pulling this switch causes the trace to trigger automatically regardless of the scope input. The slide switch marked ext, ch 1, ch 2, determines how the trace is triggered, externally through the jack directly below the switch, or triggered by the waveform being observed in channel 1 or channel 2. The ac, tv, slide switch determines trigger coupling; ac for most signals, tv for television field and line rates. The + − slide switch selects the positive or negative slope of the waveform for triggering.

Now let's push in the trigger level switch. This position, although easy to understand, is often difficult for the technician to get used to, especially if he has always used a scope with a free running trace. When there is no signal into the scope there is no trace. The trace only occurs when the signal is present. Turning the trigger level control will adjust the point on the signal waveform at which the sweep starts.

Horizontal position of the beams is controlled by the center inside position knob. Pulling this knob magnifies the horizontal time base × 5. This is a great feature for quickly inspecting small details in a waveform. The center outside control (time/division) selects the horizontal sweep speed. The left side is calibrated in milliseconds (ms.) and the right in microseconds (μs.).

Refer back to Figure 13-1B. Let's determine the exact frequency of the sawtooth waveshape. The time/division switch is set on the 20 μs./cm. position. Note that one cycle of the sawtooth takes 5 horizontal cm. to complete. Multiplying 5 × 20 μs.

gives 100 μs. for one cycle of the sawtooth. To find cycles/second convert 100 μs. to seconds and divide into 1. Here's the formula:

$$f = \frac{1}{T}$$

$$f = \frac{1}{.0001} \quad \text{(100 μs. converted to seconds)}$$

$$f = 10 \text{ KHZ}$$

Here's another example; same sawtooth waveform but now the time/division switch is set at 2 ms. The waveform takes 5 cm. to complete. So multiplying 5 × 2 ms. equals 10 ms. Plugging it into the formula gives the following answer:

$$f = \frac{1}{T}$$

$$f = \frac{1}{.01} \quad \text{(10 ms. converted to seconds)}$$

$$f = 100 \text{ HZ}$$

Channel 2 controls are identical to channel 1. Either channel's trace can be moved vertically over the CRT face. They can be interchanged as the top and bottom trace or superimposed over each other. The beauty of a dual trace scope is the fact that two related waveforms can be directly compared using the same time base. The obvious advantage of seeing and analyzing an input and output waveform simultaneously on the scope display does not have to be elaborated. The dual trace scope is indispensable for digital work. Once you become accustomed to working with a triggered sweep dual trace oscilloscope you'll wonder how you ever managed without it.

13-2. ANALYZING A SQUARE WAVE FOR TESTING AUDIO ELECTRONICS

A square wave is an ideal waveshape for testing and troubleshooting electronic circuits. When a square wave is broken down into its basic parts it consists of a sine wave and an infinite number of odd harmonics. This effect is illustrated in Figure 13-2A. Note how quickly the square wave takes shape as more

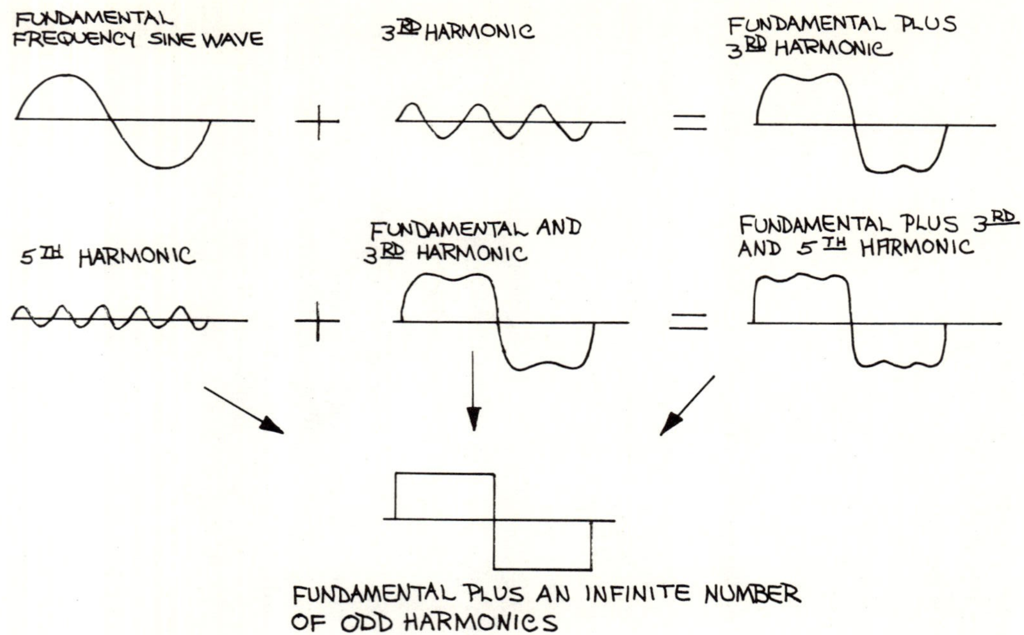

Figure 13-2A. Generation of a square wave.

odd harmonics are added to the fundamental frequency. In general terms this means that a circuit which can amplify a square wave without showing distortion is in effect amplifying the fundamental frequency and all the rest of its odd harmonics. This takes in quite a few frequencies. The beauty of it is that it's accomplished simply by injecting a square wave into the input of an amplifier and analyzing the output with an oscilloscope.

What fundamental square wave frequency should be used when evaluating the performance of audio equipment? The rule of thumb is that a square wave tests the amplifier's low frequency response down to about one tenth of its fundamental frequency. For instance, a 400 HZ square wave would test low frequency response down to 40 HZ; a 1 KHZ square wave down to 100 HZ. The place to look for a low frequency distortion is the flat top of the square wave and the trailing edge. Actually the square wave flat top is DC. Put all the flat tops together and what do you have . . . a battery, pure DC.

For checking high frequency response a square wave tests about ten times its fundamental frequency. A 1 KHZ square wave tests an amplifier up to around 10 KHZ; a 4 KHZ signal shows the response up to 40 KHZ. The leading edge of the wave is the place to look for high frequency distortion clues. A perfect square wave (no such thing) would rise from the base line to its maximum amplitude in zero time. That's a pretty fair high frequency response!

To run a test on audio equipment simply inject a square wave into the input and scope the output, Figure 13-2B. If the output is an amplified duplicate of the input, the amplifier has a flat frequency response for that particular range of frequencies represented by the square wave. Two square wave injections should do the trick for most audio apparatus. A 200 HZ square wave will test low frequency response to 20 HZ and high frequency response to about 2 KHZ. A 2 KHZ square wave will test the amplifier from 2 K and beyond the audio range. If you want to check the frequency response above the range of human hearing simply choose a higher frequency square wave.

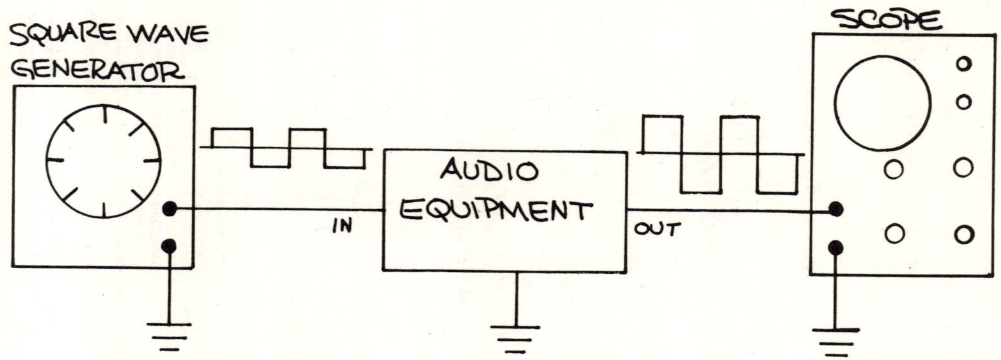

Figure 13-2B. Square wave testing audio equipment.

There are several precautions to observe when making a test of this nature:

1. The square wave can be no better than the square wave generator and the oscilloscope.
2. Apply enough input signal to drive the amplifier but not overload it. The output signal should follow any changes in input signal amplitude. A saturated or over driven amplifier will not respond to input signal amplitude changes. The correct input is that point just before the output saturates.
3. Keep all tone controls in their neutral position.
4. Substitute a noninductive resistive load for the speaker to eliminate speaker distortion. Match the resistor to the speaker in impedance and wattage characteristics.
5. Keep a good common ground between all the apparatus.

If the equipment does not respond to a square wave analysis repeat the previous test in each stage of the apparatus. The stage responsible for causing the problem will be clearly evident.

Let's take a look at a few types of distortion that show up in square waves, see Figure 13-2C. We are not going to attempt to analyze the infinite number of distortion patterns but simply show a few of the most common types.

13-3. SCOPE TECHNIQUES USING LISSAJOUS PATTERNS

A Lissajous pattern is a figure produced on the screen of an oscilloscope when sine waves of various amplitude and phase relationships are applied to the scope's horizontal and vertical deflection circuits simultaneously. These patterns can be used for frequency and phase comparisons as well as making some way-out CRT pictures. A similar technique can be used with a color bar generator and color TV receiver to produce a daisy pattern on the scope screen showing the phase angle of each color bar.

Here's the method to use when comparing a known frequency with an unknown one. Feed the reference frequency

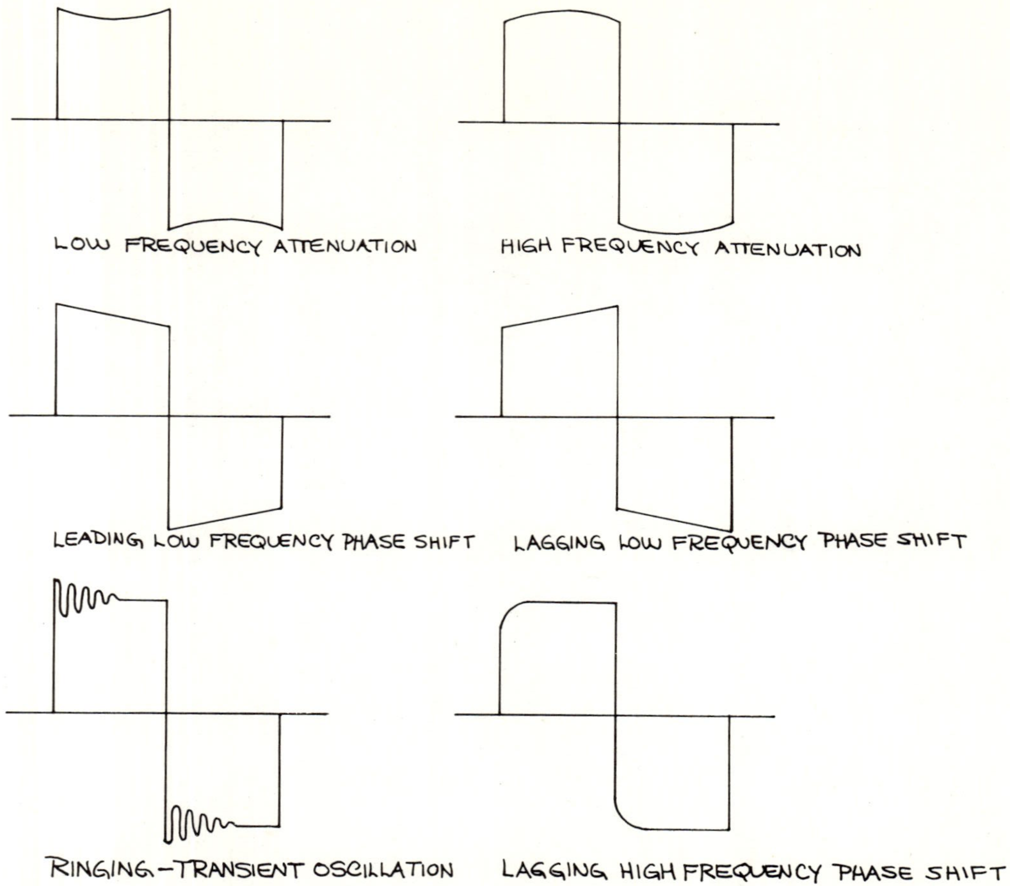

Figure 13-2C. Common square wave distortions.

(known) into the scope's horizontal input jacks. Connect the unknown frequency into the scope's vertical input jacks. Set the scope sweep control knob to the external sweep position or with a dual trace scope set the sweep for channel 2. Figure 13-3A shows typical Lissajous patterns resulting from sine wave frequency comparison.

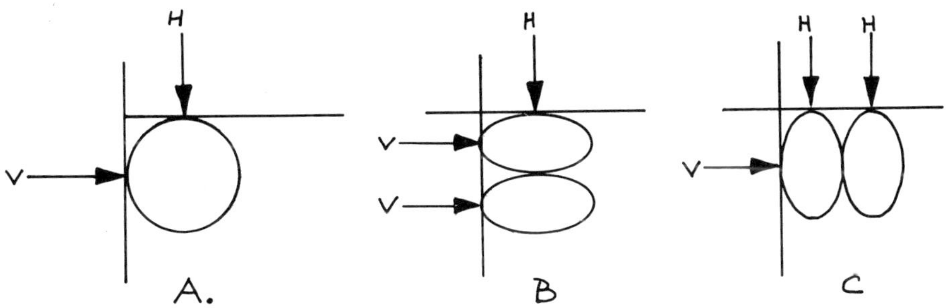

Figure 13-3A. Lissajous patterns produced by comparing two sine waves.

Notice how each Lissajous figure touches one or more points on the vertical and horizontal axis lines. These points are the secret in determining the unknown frequency. Use the following formula and rules for your calculations:

$$F_X = \frac{L_H \times F}{L_V}$$

- FX = unknown frequency
- F = known frequency
- LH = number of loops that touch horizontal axis
- LV = number of loops that touch vertical axis

Suppose the reference frequency for the patterns in Figure 13-3A is 1 KHZ. The computations for each of the figures are as follows:

A.

$$F_X = \frac{1 \times 1000}{1}$$

F_X = 1000 HZ

B.

$$F_X = \frac{1 \times 1000}{2}$$

F_X = 500 HZ

C.

$$F_X = \frac{2 \times 1000}{1}$$

F_X = 2000 HZ

The circle pattern always means the two frequencies are equal. The figure eight pattern always means the unknown frequency is one half of the known. The sideways figure eight always means the unknown frequency is twice the reference frequency. When the unknown frequency is more than ten times smaller or larger it gets pretty difficult to count the loops accurately.

Now let's use Lissajous patterns to compare phase relationships between two sine waves of the same frequency. One phase is applied to the scope's vertical input and the other phase to the horizontal input terminals. Both signals should produce equal amplitude on the scope trace. Typical phase relationships are shown in Figure 13-3B.

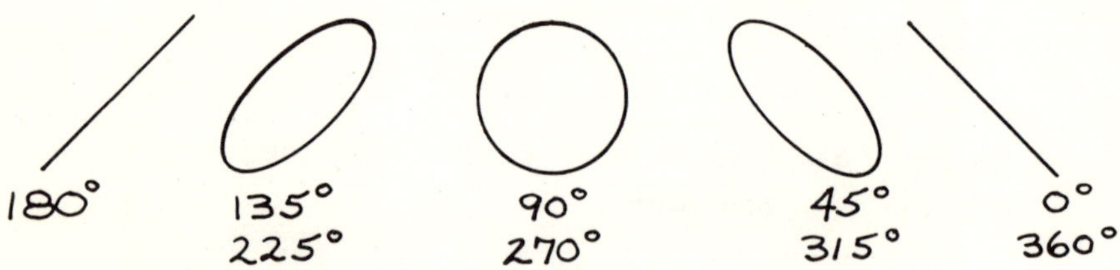

Figure 13-3B. Comparing phases between two sine waves using Lissajous patterns.

To compute the phase relationship for any shape Lissajous pattern use the formula shown in Figure 13-3C. Make sure the axis of the ellipse passes through the zero point of the graph.

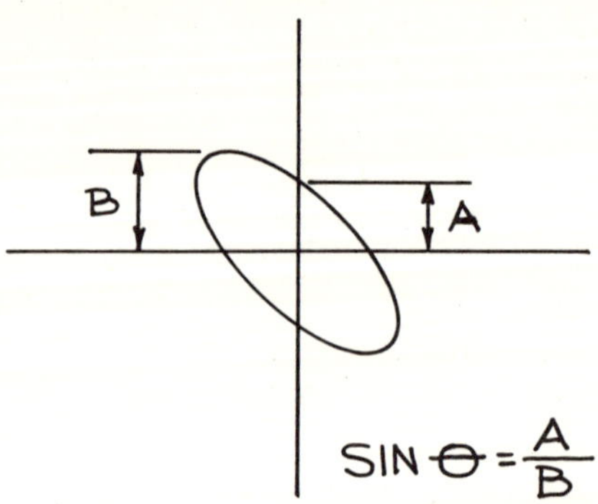

Figure 13-3C. Computing unknown phase angle between two sine waves of the same frequency.

If you want to observe the phase relationship (daisy pattern) between color bars in a color TV without a vectorscope here's all you do. Inject a color bar signal into the TV's antenna terminals. Feed the R-Y demodulator signal to the scope's vertical input jacks, and the B-Y demodulator signal to the horizontal inputs. You'll see the daisy pattern, Figure 13-3D, with each petal being a vector corresponding to the phase angle of one color bar.

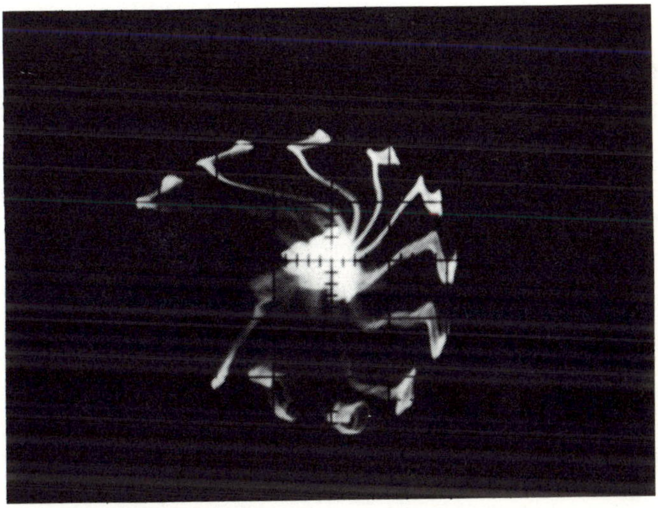

Figure 13-3D. Color TV daisy pattern.

13-4. STEPS TO UNDERSTANDING THE COMPOSITE VIDEO WAVEFORM

The complete video signal (composite video waveform) is one of the most complex electronic signals ever invented by man. Complicated as the composite video signal appears it is not hard to understand when its functions are learned pulse by pulse. It's interesting to learn about this waveform and essential for efficient troubleshooting in television circuits.

The following signals are present in the color composite video waveform:

1. B & W video
2. Color video
3. Sound
4. Vertical synchronization
5. Horizontal synchronization
6. Color synchronization
7. Vertical blanking
8. Horizontal blanking
9. Test signals

Refer to Figure 13-4A for the following discussion. The vertical blanking interval of the composite video waveform is a particularly important area to study since the remaining part of the waveform is simply repetitious. The monochrome video signals are located between the horizontal blanking pulses. Each full picture on the TV screen is produced by 525 separate video lines, see Figure 13-4B. The color video signals are also present in this same area of the composite video signal. The color information is interleaved between the B & W energy bands. The FM sound signal is also carried in this same area.

All receivers are kept in vertical synchronization with the transmitter by vertical sync pulses. Sixty of these pulses are transmitted each second. Figure 13-4C shows the vertical sync pulse broken down (serrated) into numerous smaller pulses. This is necessary to provide proper horizontal synchronization during the vertical sync time. The serrations are placed back together again in the receiver to form a single vertical sync pulse.

Horizontal sync pulses are necessary to synchronize the horizontal part of the TV screen, Figure 13-4D. One H sync pulse is necessary for each video line. There are 15,750 of them transmitted each second.

Color synchronization is provided by a special 3.58 MHZ burst signal. About eight cycles of this burst frequency are located behind each horizontal sync pulse, see Figure 13-4E.

Vertical blanking pulses are needed to shut off (blank out) the picture tube during the vertical retrace interval, Figure

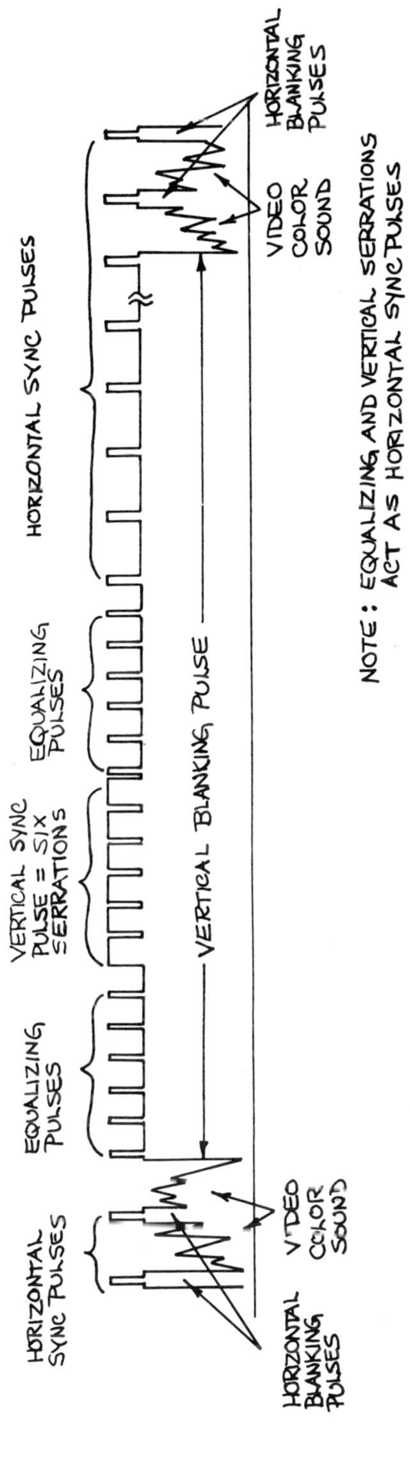

Figure 13-4A. Composite video waveform during the vertical blanking interval.

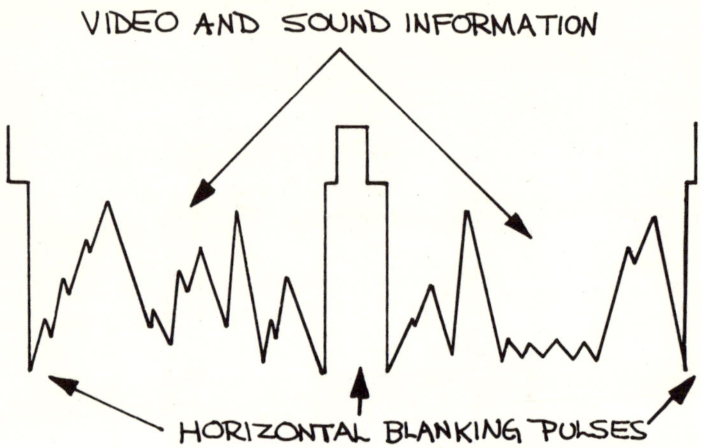

Figure 13-4B. Video and sound signals.

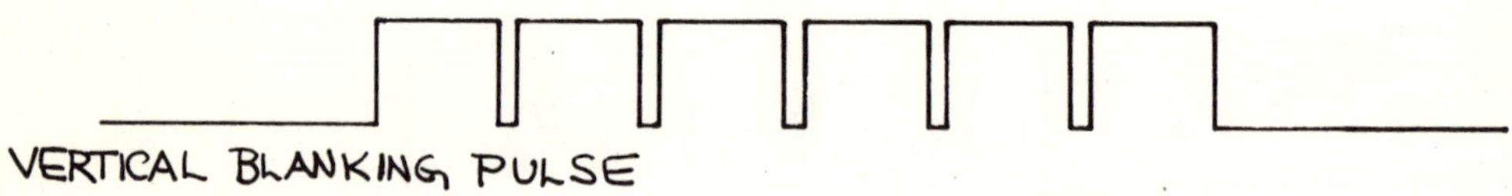

Figure 13-4C. Vertical sync pulse (serrations).

Figure 13-4D. Horizontal sync pulses.

13-4F. When the CRT scanning beam arrives at the bottom of the picture, time must be allocated for it to return to the top. During this period the vertical blanking pulse turns off the picture and the electron beam's return to the top is not seen.

There is also a need to blank out the picture tube at the end of each horizontal scanning line so the beam can retrace back to the left side of the screen to write the next line. This is accomplished by the horizontal blanking pulses, Figure 13-4G.

Test signals (VITS) are also transmitted during the vertical

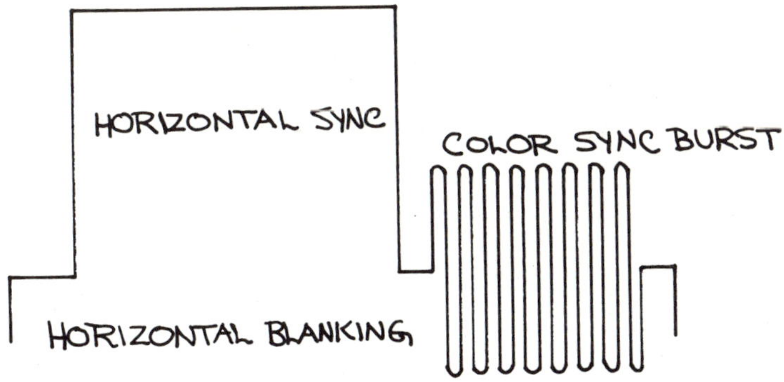

Figure 13-4E. Color sync burst.

Figure 13-4F. Vertical blanking pulse.

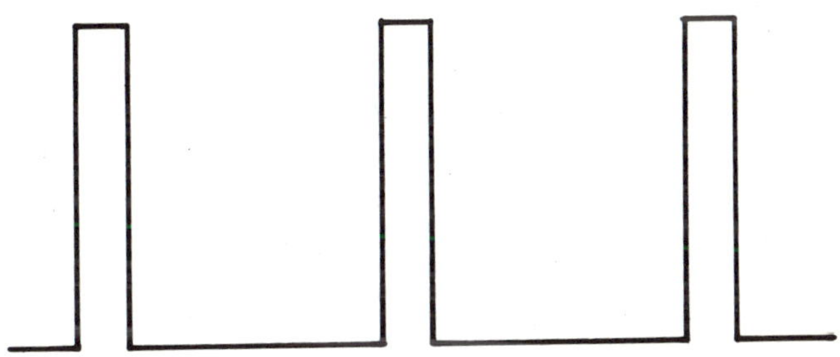

Figure 13-4G. Horizontal blanking pulses.

blanking interval, Figure 13-4H. These signals are used to provide reference values for color and luminance information and to check bandwidth.

Once the composite video waveform enters the TV receiver each part of the signal is channeled to its particular area. If any part of the composite video signal is lost or distorted in the receiver, the associated part of the picture will be affected. The following list shows what would happen to the TV picture in the absence of a particular part of the signal:

Composite Video Waveform

Missing signal	Indication on CRT
1. B & W video	No picture—screen lit
2. Color video	Good B & W picture
3. Sound	Good picture—no sound
4. Vertical sync	Picture rolls
5. Horizontal sync	Picture tears
6. Color sync	Color stripes
7. Vertical blanking	Thin white retrace lines
8. Horizontal blanking	Thin white retrace lines

A good place to observe the composite video signal in the TV receiver is at the output of the video detector. If the scope sweep is set at video line frequency (15,750 HZ) you'll see single video lines and horizontal sync and blanking pulses. When the sweep controls are set at 60 HZ/sec., with a good scope it's possible to see the waveform shown in Figure 13-4A. Trigger the scope with the TV vertical retrace blanking pulse. Don't expect to see a perfect artist's rendition of the composite video waveform. Depending on the kind of scope you're using, you just might not recognize the waveform at all. However, once you get the scope triggered properly and the controls set at their optimum position you should be able to identify all the signals and pulses described in this section. Figure 13-4I shows an actual photograph of the composite video waveform at the output of the second detector. Good luck to you!

Figure 13-4H. Vertical interval test signals.

Figure 13-4I. Photograph of composite video waveform (inverted).

Index

C

D

E

U

V

W

Z